Implementing CDISC Using SAS®
An End-to-End Guide
Second Edition

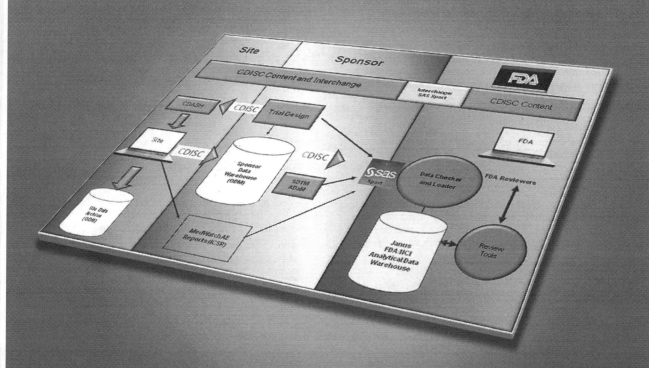

Chris Holland · Jack Shostak

The correct bibliographic citation for this manual is as follows: Holland, Chris and Jack Shostak. 2016. *Implementing CDISC Using SAS®: An End-to-End Guide, Second Edition*. Cary, NC: SAS Institute Inc.

Implementing CDISC Using SAS®: An End-to-End Guide, Second Edition

Copyright © 2016, SAS Institute Inc., Cary, NC, USA

ISBN 978-1-62959-825-3 (Hard copy)
ISBN 978-1-62960-536-4 (EPUB)
ISBN 978-1-62960-537-1 (MOBI)
ISBN 978-1-62960-538-8 (PDF)

SAS Institute Inc., SAS Campus Drive, Cary, NC 27513-2414

November 2016

Contents

About This Book

Background

The Clinical Data Interchange Standards Consortium (CDISC) started in 1997 as a global, open, multidisciplinary, non-profit organization focused on establishing standards to support the acquisition, exchange, submission, and archiving of clinical research data and metadata. The mission of CDISC is "to develop and support global, platform-independent data standards that enable information system interoperability to improve medical research and related areas of healthcare."

CDISC standards are promoted as being "vendor-neutral, platform-independent and freely available via the CDISC website." When one speaks of clinical research data, however, one particular vendor often comes to mind. SAS software has been used for analyzing, summarizing, and reporting clinical trial data since the early 1980s. The US Food and Drug Administration (FDA) has required that electronic data provided to them in marketing approval submissions be formatted in the SAS 5.0 transport file format. Although this format is now an open standard accepted by many types of software, it has played a role in establishing and maintaining SAS as the industry standard in the pharmaceutical industry for numerous tasks relating to clinical trial data.

The flagship standard developed by a CDISC working group is the Study Data Tabulation Model (SDTM). The term "tabulation" has its origins in the code of federal regulations (CFR) regarding what is required in New Drug Applications (NDAs). Specifically, 21 CFR 314.50 describes case report tabulations (CRTs) as "data on each patient in each study" that are "pertinent to a review of the drug's safety or effectiveness." Originally, these tabulations were generated as listings on paper. However, by the late 1990s, electronic data sets were playing an increasing role in new drug reviews. As a result, the SDTM was developed as a standard data format for the electronic submission of these required tabulations.

Since the inception of CDISC, its bond with SAS and SAS users in the pharmaceutical industry has been a natural marriage. Over the past decade, as CDISC standards have built up momentum in terms of stabilizing and gaining regulatory approval and preference, SAS programmers have played an increasing role in the complete life cycle of clinical trial data—from collection, to analyzing, reporting, documenting, validating, submitting and archiving data, and all the steps in between. As such, the time seems right to provide an end-to-end handbook for SAS users in the pharmaceutical industry who are working with CDISC standards.

What's New in This Edition?

When writing the first edition of this book, the FDA was still not requiring that marketing applications for new drugs and biologics be submitted electronically, much less that CDISC standards be applied to the data sets that accompanied fully electronic submissions. But the writing was on the wall, so to speak (or on the screens at industry conference presentations), that such changes were imminent. This was the motivation for that first edition—to help late adopters get on board with the someday-to-be-required CDISC implementation.

Then, in July of 2012, the FDA Safety and Innovation Act, or FDASIA, was passed, setting the wheels in motion for the requirement of electronic marketing applications including standardized study data. The Act deferred the details about the format of these soon-to-be-required electronic submissions to guidance documents, but it set a time table for them to take effect "beginning no earlier than 24 months after issuance of a final guidance." In December 2014, "Providing Regulatory Submissions in Electronic Format – Standardized Study Data" was issued as a final guidance document, thereby setting December 2016 as the due date for official conversion to CDISC standards for future submission. The guidance document, as expected, specifies the use of both SDTM and ADaM datasets.

But these requirements by themselves were not uniquely motivating for a new edition of our book. As we were writing the first edition, we knew that the rapid evolution of CDISC standards would render at least part of the book outdated before it even came out. Perhaps most notable among these changes was the release and finalization of the Define-XML 2.0 standard, a more extensible version of the original standard for dataset meta-data. Many other changes have also since been released, including a host of new SDTM domains, a new version to the ADaM Implementation Guide (1.2), a new data structure for occurrence-related analysis data sets (the OCCDS), new ADaM validation checks (now on version 1.3), new Pinnacle 21 software, new therapeutic area standards, and, of course, new versions to the SAS and JMP software used to implement these standards. Many of these updates have been reflected within their respective relevant section within the book.

What hasn't changed is the process by which we recommend that these standards be implemented. The use of CDASH at the collection stage is more important than ever, although mostly outside the scope of this book. The subsequent "linear" conversions from the raw, collected data to the SDTM and ADaM standards are still very much a basic tenet of that process. We also continue to espouse the virtues of establishing the metadata before the creation of the actual data that they describe. So from that perspective, the "end-to-end" organization of the book has remained largely unchanged.

Is This Book for You?

Any manager or user of clinical trial data in this day and age is likely to benefit from knowing how to either put data into a CDISC standard or analyze and find data once it is in a CDISC format. If you are one such person—a data manager, clinical and/or statistical programmer, biostatistician, or even a clinician, then this book is for you.

Prerequisites

It is assumed that the readers of this book have some basic understanding of SAS programming. It is also advantageous if the reader has some familiarity with the CDISC SDTM and ADaM models. If you do not have this knowledge, then there will be some portions of this book that will be more difficult to follow.

Purpose

This book has two purposes. One is to introduce readers to the end-to-end process for implementing CDISC standards on clinical trial data. The second is to introduce readers to tools based on SAS that cannot only facilitate the implementation process but also facilitate the ultimate goal of analyzing your data once it follows a CDISC standard. Some of these tools are actual SAS products. Some are SAS macros provided by the authors.

In order to achieve these goals, the authors have created fictional clinical trial data that will be used to provide examples of how to apply SAS tools to the various steps in the process along the way.

Organization and Scope of This Book

With the end-to-end concept in mind, this book is organized with a start-to-finish mentality for clinical trial data. Chapter 1 starts with an overview of the primary standards covered in this book, the SDTM and the Analysis Data Model (ADaM), and a discussion of considerations as to when each should be implemented. Chapter 2 is focused on what should be the first step to any big project, the specifications. Although many currently think of the creation of metadata and data documentation as part of the process that comes after building the datasets, the authors hope that the tools and tips provided here will facilitate the creation of a define.xml file as a tool for implementing SDTM standards before any of the actual conversion work is performed.

Chapters 3, 4, and 5 all cover SDTM implementation, but with different tools (Base SAS, SAS Clinical Standards Toolkit, and SAS Clinical Data Integration, respectively). Chapter 6 relates to ADaM metadata while Chapter 7 covers ADaM implementations using Base SAS. With regulatory, quality, and compliance concerns in mind, Chapters 8 and 9 cover CDISC validation using SAS software and Pinnacle 21 software, respectively. Chapter 10 is about analyzing data that have been structured to follow CDISC standards, with a particular focus on JMP Clinical, which has been specifically developed to work on SDTM and ADaM data. Finally, concluding the "end-to-end" process, Chapter 11 provides information on how to integrate your data from various studies into one integrated database and other steps required for providing your data in a regulatory submission.

Although many CDISC standards have stabilized, there are new standards on the horizon. Chapter 12 looks at some of these uncertainties, such as the Janus clinical trial data repository and HL7 messaging for CDISC content. Other topics relating to CDISC standards are covered here as well, such as the SEND model for non-clinical data and the BRIDG model for relating HL7, CDISC, and other standards, models, and semantics together.

What is not covered in this book are details, guidelines, and examples on how to put certain types of data into the CDISC models. These are details that CDISC therapeutic area working teams are most adept at dealing with. The SDTM implementation guide (IG), contains a wealth of knowledge pertaining to how, when, and where to put clinical trial data in SDTM domains. Similarly, the ADaM IG and associated appendixes have more information to help implementers decide how to build ADaM datasets.

Software Used to Develop This Book's Content

As you might expect, the majority of the software used in this book is SAS software. The following software was used in the production of this book:

- Base SAS 9.4
- SAS Clinical Standards Toolkit 1.6
- SAS Clinical Data Integration 2.6
- JMP 11.2
- JMP Clinical 5.0
- Pinnacle 21 Community 2.1.0

Various versions of Internet Explorer and Chrome were used to render the define.xml files

Data and Programs Used in This Book

The majority of the source data for this book can be generated by the SAS programs found in Appendix A. There is some study metadata used to generate the Trial Design Model datasets for the SDTM that you can find in Microsoft Excel files on the author pages listed below. The metadata spreadsheets used for the ADaM and SDTM data can also be found on the author pages listed below. Finally, all programs used in this book can be found on the author pages as well.

Author Pages

You can access this book's author pages at:

http://support.sas.com/hollandc

http://support.sas.com/shostak

The author pages include all of the SAS Press books that these authors have written. The links below the cover image will take you to a free chapter, example code and data, reviews, updates, and more.

Example Code and Data

You can access the example code and data for this book from the author pages listed in the previous "Author Pages" section. On the author page, select "Example Code and Data" to display the SAS programs that are included in this book.

For an alphabetical listing of all books for which example code and data are available, see http://support.sas.com/documentation/onlinedoc/code.samples.html. Select a title to display the book's example code.

If you are unable to access the code through the website, send email to saspress@sas.com.

Additional Help

Although this book illustrates many analyses regularly performed in businesses across industries, questions specific to your aims and issues may arise. To fully support you, SAS Institute and SAS Press offer you the following help resources:

- For questions about topics covered in this book, contact the author through SAS Press:
  - ○ Send questions by email to saspress@sas.com; include the book title in your correspondence.
  - ○ Submit feedback on the author's page at http://support.sas.com/author_feedback.
- For questions about topics in or beyond the scope of this book, post queries to the relevant SAS Support Communities at https://communities.sas.com/welcome.
- SAS Institute maintains a comprehensive website with up-to-date information. One page that is particularly useful to both the novice and the seasoned SAS user is its Knowledge Base. Search for relevant notes in the "Samples and SAS Notes" section of the Knowledge Base at http://support.sas.com/resources.
- Registered SAS users or their organizations can access SAS Customer Support at http://support.sas.com. Here you can pose specific questions to SAS Customer Support; under *Support*, click *Submit a Problem*. You will need to provide an email address to which replies can be sent, identify your organization, and provide a customer site number or license information. This information can be found in your SAS logs.

Keep in Touch

We look forward to hearing from you. We invite questions, comments, and concerns. If you want to contact us about a specific book, please include the book title in your correspondence.

Contact the Author through SAS Press

- By e-mail: saspress@sas.com
- Via the Web: http://support.sas.com/author_feedback

Purchase SAS Books

For a complete list of books available through SAS, visit sas.com/store/books.

- Phone: 1-800-727-0025
- E-mail: sasbook@sas.com

Subscribe to the SAS Learning Report

Receive up-to-date information about SAS training, certification, and publications via email by subscribing to the SAS Learning Report monthly eNewsletter. Read the archives and subscribe today at http://support.sas.com/community/newsletters/training!

Publish with SAS

SAS is recruiting authors! Are you interested in writing a book? Visit http://support.sas.com/saspress for more information.

Acknowledgments

This book would not have been possible without the thorough review and thoughtful feedback of several individuals that we would like to thank and recognize for their contribution.

Internal SAS support:

- Editor: Stacey Hamilton
- Technical Publishing Specialist: Denise T. Jones
- Cover design: Robert Harris
- Copyeditor: Kathy Underwood (with marketing materials edited by John West)

Technical reviewers:

- Lex Jansen
- Nate Freimark
- Bill J. Gibson
- Geoffrey Mann

Jack would also like to dedicate this edition in memory of his father John and dog Susie. They taught him many lessons in how to do the right things and how to smell the roses, respectively.

Chris would like to dedicate this edition to all of the devoted CDISC volunteers who contribute so much of their time and energy to the worthy cause of developing data standards in our industry.

About These Authors

CHRIS HOLLAND has been a SAS user since 1990. He currently works as a Biostatistics Director for Amgen and has prior experience heading up biostatistics departments for Sucampo Pharmaceuticals, MacroGenics, and Micromet. He has also worked as a statistical reviewer at the Center for Drug Evaluation and Research in the U.S. Food and Drug Administration. It was there where he served as the technical lead for the SDTM/ADaM Pilot Project FDA review team and where he invented an early version of the MAED Service, an adverse event review tool that is currently in production at the FDA. Holland continues to be active in the CDISC community, particularly with the ADaM team. He received an MS in statistics from the University of Virginia, a BS in statistics from Virginia Tech, and is an Accredited Professional Statistician™ by the American Statistical Association.

JACK SHOSTAK, Associate Director of Statistics, manages a group of statistical programmers at the Duke Clinical Research Institute. A SAS user since 1985, he is the author of *SAS® Programming in the Pharmaceutical Industry, Second Edition* and coauthor of *Common Statistical Methods for Clinical Research with SAS® Examples, Third Edition*, as well as *Implementing CDISC Using SAS®: An End-to-End Guide*. Shostak has published papers and given talks for various industry groups, including the Pharmaceutical SAS Users Group (PharmaSUG) and PhUSE. He is active in the Clinical Data Interchange Standards Consortium (CDISC) community, contributing to the development of the Analysis Data Model (ADaM), and he serves as an ADaM trainer for CDISC. Shostak received an MBA from James Madison University and a BS in statistics from Virginia Tech.

Learn more about these authors by visiting their author pages, where you can download free book excerpts, access example code and data, read the latest reviews, get updates, and more:
http://support.sas.com/publishing/authors/holland_chris.html

http://support.sas.com/publishing/authors/shostak.html

Chapter 1: Implementation Strategies

The Case for Standards

The decision to adapt to CDISC standards within an organization or for a particular clinical development program has gotten easier since Congress approved the FDA Safety and Innovation Act, or FDASIA, in July of 2012. As of December 2016, the implementation of CDISC standards, primarily the Study Data Tabulation Model (SDTM) and Analysis Data Model (ADaM), is required for certain studies contained within FDA marketing applications. For the early adapters, this change will have no impact on their current processes. These are the organizations that clearly saw the benefits of adapting to CDISC as soon as possible. Mergers and acquisitions have persisted throughout the pharmaceutical industry for decades, and behind the scenes of each merger are the biostatisticians, data managers, and SAS programmers who've worked at the same desk year after year, but have seen their employer names change three to four times. Throughout it all, with a change in the employer came the change in the case report form

(CRF) designs, variable names, and data formats for the different compounds on which they worked. When it came time to integrate the data for a regulatory submission, a substantial amount of time was spent deciding on the structure and variable names to be used for the integrated database. And that was just the beginning. The time spent doing the actual conversions and integration is often much greater. As the programming hours piled up, those involved started to see the merits of having a standard across the industry.

Pharmaceutical and biotech companies weren't the only organizations undergoing mergers. During the late 1990s and early 2000s, many contract research organizations (CROs) consolidated as well. In addition to the numerous data standards that they had to keep track of among their various clients, CRO SAS programmers also had to deal with different data formats being used internally due to consolidation with other CROs. Some at these CROs got to work on integration projects involving compounds that, at each new phase of development, had been passed from one organization and CRO to another. As a result, even the most basic key identifier of any clinical trial dataset, the subject ID, was sometimes uniquely named within each study. So as the programming hours piled up, the key decision makers at CROs started to see the merits of having a data standard across the industry.

Yet this grass-roots initiative to develop industry-wide standards would not have gotten off the ground without the support of the biggest consumer of clinical trial data of all, the US Food and Drug Administration. For some time, FDA reviewers had to deal with completely different data formats and structures from one sponsor to the next. This might not have been so cumbersome in the days before the Prescription Drug User Fee Act (PDUFA, commonly pronounced *puh-DOO-fa*) first became effective. Before PDUFA, a review clock was non-existent, and two-year reviews of New Drug Applications (NDAs) and Biologic License Applications (BLAs) were the norm. However, with the passage of PDUFA, review cycles were originally mandated to be 12 months (and are now down to 10 months). With those review clocks, along with increasing expectations to carefully inspect the electronic data that were packaged with NDA and BLA submissions, reviewers found themselves having to do more with less.

The aftermath of some pivotal events in 2004 put even more pressure on FDA reviewers. One was the investigation of suicidality risks among children on antidepressants. The other was the withdrawal of Vioxx from the market. Because of these two high-profile safety concerns, doctors, patients, and sponsors all suddenly had a vested interest in knowing whether the drugs that they were prescribing, taking, or selling to treat depression, arthritis, or any number of spin-off indications were adding risks that outweighed the benefits. The brunt of these class-effect determinations fell on the FDA clinical and statistical reviewers who were the only ones who had access to all the clinical trial data that would allow them to make the informed decisions that the public, doctors, industry, congressmen, and media were all suddenly demanding. However, when the drug class under consideration involved 10 different compounds from as many different sponsors with as many different data formats, this was no easy task. "Wouldn't it be great," some FDA reviewers asked, "if we had all the data we need in one giant database?" Fortunately, within the FDA, certain reviewers, team leaders, and division directors all started to see the merits of having a data standard across the industry. To coin a common phrase of one particular FDA division director who had a penchant for promoting data standards at industry conferences, the mantra of the late 2000s became "just CDISC-It."

Evidence of FDA's support of data standards is not only found at conference podiums. Since the draft release of version 3.1 of the SDTM Implementation Guide (IG) in October 2003, the FDA has issued a number of documents indicating their support of data standards. A summary of these appears in Table 1.1.

Table 1.1: FDA Documents and Events in Support of Data Standards

Time	Event
July 2004	eCTD study data specifications reference the SDTM for tabulation data
March 2006	"Development of data standards" listed as opportunity #44 in the FDA's Critical Path opportunities list
September 2006	SDTM/ADaM pilot project review completed and results presented at the CDISC Interchange
September 2006	Old e-NDA guidance document is withdrawn (leaving the eCTD study data specifications as the only guidance relating to submission data)
December 2006	Proposed rule to require electronic data with submissions is released in the Federal Register
May 2008	First version of the PDUFA IV IT plan is released, making numerous commitments to the SDTM
October 2009	Version 1.5 of the study data specifications released, making specific reference to the Analysis Data Model (ADaM) standard for analysis data
March 2010	Version 1.0 of the CDER Data Standards Plan released, providing a commitment to CDISC standards
May 2011	The "CDER Common Data Standards Issues Document," version 1.0 is released, stating that CDER is "strongly encouraging sponsors to submit data in standard form"
February 2012	Draft FDA guidance document *Providing Regulatory Submissions in Electronic Format—Standardized Study Data* (a.k.a. "the e-Data Guidance") released to establish "FDA's recommendation that sponsors and applicants submit study data in a standardized electronic format"
December 2014	Final FDA e-Data guidance. Started the 24-month clock for requiring NDAs, aNDAs, and certain BLAs and INDs to be made available electronically to CDER and CBER and in the required format as specified by the Data Standards Catalog (available at http://www.fda.gov/forindustry/datastandards/studydatastandards/default.htm)

Nowadays, the legion of CDISC implementers is tangible to any SAS user conference attendee who struggles to find an empty chair in a session that has anything to do with CDISC. Managers are preaching the data standards gospel, software vendors are demonstrating their tools that use CDISC data, FDA presenters are promoting their preference for CDISC, and FDA documents are requiring the implementation of SDTM and ADaM models for sponsors' NDA and BLA submissions. The question is no longer *whether* to implement CDISC standards. Rather it is now more a question of *when*, and how far to go with it.

Which Models to Use and Where

The advantages of having universal data standards are largely geared toward users of the data for review or analysis. Across studies, medical and statistical data reviewers and analysts, whether they are on the sponsor side of the equation or on the regulatory review side, benefit from having nearly instant familiarity with how data are organized for any given study. This holds true whether the data are non-clinical SEND data, SDTM tabulation data, or ADaM analysis files.

However, for those who are responsible for putting the data in these standardized formats, there is much more work involved. Before data standards, there was often just one set of data provided with NDA and BLA submissions. Many of these datasets tended to be a hybrid between the raw CRF data and the analysis files. The structures and variable names often matched those of the original database tables and therefore required little manipulation. Now, not only do you have to worry about a potentially labor-intensive conversion process from the raw data tables to the SDTM domains, you also need to then create ADaM datasets from the SDTM domains.

For organizations with plenty of resources to devote to the implementation of standards, this process might be manageable. CROs who conduct a high volume of conversions for their clients have an opportunity to streamline their implementation process with each new iteration. Certain technologically advanced organizations such as software companies with proprietary electronic data capture (EDC) systems and expert knowledge of the data standards are capable of developing automated tools to assist with the conversion process from the raw CRF data to fully compliant SDTM domains and subsequent ADaM data files.

Many large organizations with high volume and adequate resources are now implementing CDISC standards as early as Phase 1, despite the historically low chance of a drug ultimately advancing to the marketing submission stage. For others, the wait-and-see approach might be more appealing, given their lack of expertise and resources.

Eventually, however, certain medical products will advance in development, and when they do, it is better to be prepared. As such, the objectives of this book are to provide the following:

- Considerations for deciding when to start implementing CDISC standards
- Advice for how to get started with CDISC implementation and how to move forward with it
- An introduction to SAS software tools that assist with the creation of CDISC data and metadata documentation (and instructions on how to use them)
- Information on how to check CDISC data for compliance
- Information about tools for using CDISC data for analysis and review

Starting with the Clinical Data Acquisition Standards Harmonization (CDASH) Standard

The best way to adapt to being a CDISC organization is to start implementing standards at the initial step of data acquisition—the CRFs. The Clinical Data Acquisition Standards Harmonization (CDASH) (http://www.cdisc.org/standards/foundational/cdash) standard was created in response to opportunity #45 in FDA's Critical Path Opportunities list, which was titled "Consensus on Standards for Case Report Forms." Although part of the initiative was to standardize the look and feel of CRFs, a big part of the initiative in the eyes of CDISC implementers was to standardize the variable names of data elements being captured in the clinical database. Having such a standard that was consistent with SDTM terminology would make the conversion to SDTM much easier. With a CDASH structure behind any data management system, certain SDTM domains, like adverse events (AE), demographics (DM), and concomitant medications (CM), are almost instantly SDTM-ready with the initial data extract to SAS datasets. In total, 16 domains are covered by the CDASH standard, covering those that are common to most therapeutic areas and types of clinical research. For further reading, the CDASH-published document (version 1.1 was published in January 2011) contains implementation recommendations, best practices, regulatory references, and loads of other information pertinent to CDASH.

For any organization starting from the ground up, implementing CDASH should be an easy decision because it precludes the need to develop a new organization-specific standard. However, unless the data management system happens to come pre-packaged with CDASH default templates, implementing an existing standard can still require considerable work. Without these templates, one important element to a successful implementation is making sure that the proper know-how is put to work. Just a basic knowledge of CDASH might not be enough. Having a breadth of knowledge that spans CDASH, SDTM, and ADaM can help prevent you from, for example, having variable names in the source data that conflict with variables in the SDTM or ADaM data. A careful deployment with proper attention to downstream standards can save you from unnecessary variable renaming later on.

Whatever the situation, whether the true source data is from an entirely CDASH environment or from something that resembles nothing of the sort, the source data can be considered just various shades of gray in the eyes of an SDTM implementer. Before delving into the programmatic conversion process, the very important step of mapping out a conversion plan needs to be discussed.

Implementation Plans and the Need for Governance

Before an actual CDISC implementation takes place, whether it is a conversion from CDASH to SDTM or the creation of ADaM data from the SDTM, it is often a good idea to document the precise mapping from one data source to another. The advantages of this are three-fold:

- It allows the work to be handed off from the planner(s) to programmer(s), thereby obviating the need to have these functions performed by the same individual.
- It provides a plan to the programmers that has been discussed, reviewed, and approved ahead of time. It will also prevent ad hoc decisions by one programmer conflicting with those of another on the same project.
- It provides a specification that the final work product can be checked against and referred to along the way.

Anyone who has spent much time trying to implement a CDISC standard has probably quickly realized that, despite efforts to the contrary, much of it is subject to interpretation. Consequently, there is a strong likelihood that one person's interpretation is different from another's. And herein lies the foundation for another form of conflict relating to standards—the friction between two or more strong-minded individuals who each have their own opinion on how to implement.

In order to handle this inevitable problem, many organizations have developed a form of governance where decisions relating to controversial issues are agreed upon by a group of experts. The process by which these issues are presented to and decisions are made by a governing board can vary. Either the board can be responsible for reviewing and approving all document specifications developed within an organization, or they can only get involved to weigh in on certain issues, especially the overarching ones that are likely to affect all projects.

For smaller organizations, use of a governing board might be unnecessary or impractical. Mapping decisions can be made either by senior personnel or by outside consultants. Whatever the size or status of an organization, in order to avoid conflicts later on, reviewing and approving mapping specifications before the actual work begins can, at the very least, prevent bad decisions from being made simply because they reflect work that has already been done.

SDTM Considerations

As mentioned earlier, the decision about how and when to implement the SDTM is not always an easy one. Waiting until a Phase III study is unblinded and a pre-NDA meeting occurs can often mean having to convert a lot of data in a short amount of time. On the other hand, converting all studies, starting with the first-in-man Phase I, can mean spending a lot of effort on conversions for studies that might never even get into late Phase II or Phase III trials, where the benefits of SDTM conversions can really pay off.

Organizations struggling with these decisions should consider the following questions:

- Do you have the proper expertise and resources to implement the SDTM?

    Proper and compliant implementation is important in order to ensure that tools that depend on standards work properly and that users of your data (such as regulatory reviewers) have a pleasant experience doing so. Although the objective of this book is to help make the process easier, it will not teach the subtle details of the SDTM. The best reference for that is the most recent version of the SDTMIG. It is full of details and instructions that should not be overlooked. For trickier problems, such as how to model data that don't seem to have an explicit domain, seek advice from consultants or online experts on any of the various message boards available. But even with the proper expertise, the conversion process can be a tedious one. Make sure that you have sufficient resources to conduct a proper implementation.

- Do you have enough studies in the pipeline that would allow for an efficient and steep learning curve if every study were to be converted?

    Like everything, practice makes perfect, and the less time you spend between implementations, the less you tend to forget how things were done the last time around. As such, a one-off SDTM conversion will not allow you to fine-tune the process with subsequent iterations.

- Do you have a stable environment to allow automation of certain parts of the conversion process from study to study?

    Foundational changes, such as corporate mergers or your EDC vendor going out of business, are difficult to prepare for. In some situations, however, you might be able to anticipate the likelihood of having different database designs across studies. If the designs do change, then you'll have trouble building an automated conversion processes across studies. The first conversion will be a learning experience regardless. But with each subsequent conversion, the more similarities there are with the raw CRF data across studies, the more opportunities you will find to make the conversion more efficient, such as using SAS macros or standard programs for converting certain domains.

- Do you plan on using any tools that could make downstream processes such as data cleaning, safety reviews, or analysis more efficient when used on SDTM data?

  Certain off-the-shelf tools can make data review, particularly safety data review, easier if the data are SDTM-compliant. If you would like to produce patient profiles or other reports and summaries with review tools that leverage the SDTM, then you will certainly benefit from an SDTM conversion. Some of these review tools will be discussed in this book.

- What phase of development are you in?

  Many regulatory guidance documents provide advice about how to incorporate safety data into a submission. They tend to differentiate between Phase I safety data from healthy volunteers and those from Phase II and III studies that are more pertinent to the population, dose, and treatment regimen being considered for approval. You must also consider the attrition rate of experimental therapies. Products that eventually make it to a regulatory submission are the exception rather than the norm. And when or if they do make it to submission, not integrating or converting the Phase I data might be an option to consider because such data, aside from potential PK results, PD results, or both, are less relevant to a review of a product's safety and efficacy. Products at later stages of development, however, might reap better rewards as a starting point for implementing the SDTM.

- Should I consider a staged approach?

  Perhaps you or your organization lacks the resources or expertise for a full-blown SDTM conversion. You might still benefit from having certain key domains, such as adverse events, demographics, concomitant medications, and laboratory data in a format that, if not fully SDTM-compliant, is pretty close. Doing so will facilitate the development of standard programs, might be sufficient to use certain tools, and will make a full conversion, if required later on, that much easier. However, keep in mind that an FDA submission will likely require a fully compliant implementation.

ADaM Considerations

The first version of the ADaM model document was released in final form in December of 2004. It contained general considerations with respect to analysis datasets. Starting in April 2006, the ADaM team began working toward two significant goals:

- To define a standard data structure that would work well for many common analysis needs
- To create an ADaM Implementation Guide

Around the same time, the idea for a mock submission that FDA reviewers could use to see how well both the SDTM and ADaM data standards met their needs for a mock review started to gain some traction. This idea developed into the first SDTM/ADaM pilot project. During the course of a year, volunteers from industry worked feverishly to get this sample submission together, and volunteers from the FDA worked diligently to closely evaluate the data, compile their comments, and discuss their findings. The constructive feedback assisted the ADaM team in its work on a new version of the model document and the first-ever implementation guide.

Drafts of the model document (ADaM version 2.1) and the implementation guide (ADaM IG version 1.0) were posted for public comment in May 2008. Final versions of both documents were published in December 2009 and serve as the basis for topics relating to ADaM in this book. They can be found on the CDISC website at http://www.cdisc.org/standards/foundational/adam. Although updated final versions to these models will be available by the time this edition is published, they will not impact this book.

The ADaM 2.1 model document highlights certain fundamental principles relating to ADaM data. As stated in the document, analysis datasets and their associated metadata must have the following characteristics:

- Facilitate clear and unambiguous communication
- Provide traceability between the analysis data and its source data (ultimately SDTM)
- Be readily usable by commonly available software tools

Further, analysis datasets must have the following characteristics:

- Be accompanied by metadata
- Be analysis-ready

The decision about whether to implement ADaM standards within an organization should be an easier and more straightforward one compared to the SDTM decision. First of all, an assumption with ADaM data is that there is corresponding SDTM data to which your ADaM data can be traced. So, the first question you have to ask is whether SDTM source data exists. If so, the next question is how extensive a study, from an analysis perspective, are you dealing with. The effort to create ADaM data from SDTM data for small, safety- and PK-based Phase I studies should be balanced against a potentially limited benefit. This is because a) analyses for such studies are usually quite basic and b) analysis datasets from such trials are rarely expected for a regulatory submission. It is highly recommended, however, that at least the ADSL dataset be created. This one dataset, which includes just one record per subject, is the minimum dataset requirement for an ADaM submission. Even for small, early-phase trials, it can be useful as a single source for capturing flags and certain key information relating to each subject's experience in the trial.

Not all analysis data will fit into one of the predefined analysis data structures, such as the ADaM Basic Data Structure (BDS). If you are *not* using the specific data structures mentioned in ADaM documents, you should at least consider following the basic principles for analysis datasets mentioned previously.

The authors hope that the tools based on SAS software covered in this book and the information on how to use these tools will make conversions to CDISC standards easy enough to balance out with the efficiencies gained by using CDISC data for exploration and analysis.

Chapter Summary

The motivations for having industry-wide data standards are multi-fold. The decision about when and how to adapt to any CDISC standard is, however, a bit more complicated. In this chapter, we presented some considerations to assist with these decisions. In this book, we will be providing many tools and examples to make the conversion process easier.

Chapter 2: SDTM Metadata and Define.xml for Base SAS Implementation

This chapter provides the SDTM metadata required to implement our Base SAS solution, which creates an SDTM database. Microsoft Excel files are used to store the SDTM metadata that the SAS code (presented in Chapter 3) will leverage to create the SDTM datasets. The chapter ends with a call to the SAS macro using the SDTM metadata that is stored in Microsoft Excel to create the define.xml file.

More Information

Appendix B - SDTM Metadata

Author Pages (http://support.sas.com/publishing/authors/index.html) - %make_define2 SAS Macro

See Also

Study Data Tabulation Model Implementation Guide version 3.2

Define-XML version 2.0

Study Data Technical Conformance Guide

SDTM Metadata

It is our assertion that in order to build SDTM files quickly and accurately, it is essential to have good metadata to drive your process. However, for the sake of argument, let us look at just using Base SAS alone to perform the SDTM data conversions. You might write SAS programs that look something like this:

```
data dm;
  set sourcedata.demographics;
  length usubjid $ 24 race $ 30;        ❶
  label usubjid = "Unique Subject ID"   ❷
        race   = "Race";
  usubjid = put(subject,10.);
  if nrace = 1 then       ❸
    race = 'WHITE';
  else if nrace = 2 then
    race = 'BLACK OR AFRICAN AMERICAN';
  else if nrace = 3 then
    race = 'OTHER';
run;
```

❶ Variable lengths are buried in the SAS code, which makes them hard to change globally and difficult to make consistent across domains.

❷ Variable labels are buried in the SAS code, which makes them hard to change globally and difficult to make consistent across domains.

❸ SDTM-controlled terminology is buried in the SAS code, making it hard to verify, maintain, and get into the define.xml file.

If you take this type of Base SAS approach to your SDTM conversion work, then you will run into the following problems because your SDTM metadata is buried in your SAS code:

- Your SAS code will be hard to maintain.

- Achieving variable attribute consistency across SDTM domains will be difficult. Lengths, labels, and variable types can vary for the same variable across domains.

- You will have a very difficult time building a proper define.xml file for your submission. At best, your define.xml file will be based on the data that was observed and will omit possible controlled terminology values that are not actually observed.

In this chapter, we use Microsoft Excel files to house our SDTM metadata for later use in SAS. You could, and probably should, have your metadata stored in a more robust tool than Microsoft Excel files, such as a relational database such as Oracle or a tool like SAS Clinical Data Integration (to be discussed in Chapter 5). However, for the purposes of this chapter, we chose the Excel file format because it is readily available to most users and tends to be the common exchange mechanism of business data in lieu of sufficient software. Our goal is to have a central SDTM metadata repository so that you can more easily generate valid SDTM domains and so that you can build a proper define.xml file in the end.

The CDISC metadata standard that we comply with is the Define-XMLfinal version 2.0 from the http://www.cdisc.org/define-xml website, published in March of 2013. The SDTM target metadata that we use is based on the SDTM version 1.4 and Implementation Guide version 3.2. To view our define.xml file, we will use the define2-0-0.xsl style sheet that comes with the

Define-XML final version 2.0 release package, which can also be found on the author pages. The CDISC-supplied metadata spreadsheet for the SDTM, which can be found on the CDISC SHARE website at http://www.cdisc.org/standards/share, provides a good source for populating some of the SDTM metadata mentioned later as well. Unfortunately, this metadata spreadsheet is currently available for download only for CDISC members.

Table of Contents Metadata

The Table of Contents metadata is called that because it represents the metadata required to produce the Table of Contents section of the define.xml file. The Table of Contents in the define.xml file is just what it sounds like, a list of the domain datasets, what they are for, and where they can be found. This metadata is used to populate the ItemGroupDef section of the define.xml file.

The metadata that we use in the Table of Contents metadata is described in the following table.

Table 2.1: Table of Contents Metadata

Excel Column	Description
OID	The unique ODM element identifier (OID) of the SDTM domain and a merge key with the variable level metadata. This can be the same as NAME but does not have to be. OIDs are unique ODM identifiers and do not have to match end-user NAMEs. OIDs enable define.xml elements to reference one another.
NAME	The name of the SDTM domain that will be placed in define.xml.
REPEATING	Set to `No` for domains that are one record per subject like DM. Set to `Yes` for domains that have more than one record per subject.
ISREFERENCEDATA	Set to `No` for subject-level data. Set to `Yes` for reference data such as the trial design model datasets.
PURPOSE	Purpose for SDTM datasets should be set to `Tabulation`.
LABEL	Descriptive label of the domain. You can find these defined for the standard domains in the CDISC SDTM documentation.
STRUCTURE	Put domain dataset structure here. You can find these defined for the standard domains in the CDISC SDTM documentation.
CLASS	Set to the general SDTM class. You can find these defined for the standard domains in the CDISC SDTM documentation.

Excel Column	Description
ARCHIVELOCATIONID	This is the file reference to where the actual domain dataset lives. If you assume that the SDTM domains will be located in the same folder as the define.xml file, you can just set this to ./domainname where *domainname* is your domain name (for example, ./AE).
COMMENTOID	Set to a COMMENTOID value that should exist in the COMMENTS spreadsheet.

To see the Table of Contents metadata that we use for this chapter, refer to "Appendix B.1 – Table of Contents Metadata." When the Table of Contents metadata is put in the define.xml and rendered in a web browser with the stylesheet, it looks like this:

Tabulation Datasets for Study XYZ123 (SDTM-IG 3.2)

Dataset	Description	Class	Structure	Purpose	Keys	Location	Documentation
TA	Trial Arms	Trial Design	One record per planned Element per Arm	Tabulation	STUDYID, ARMCD, TAETORD	ta.xpt	
TD	Trial Disease Assessments	Trial Design	One record per planned schedule of assessments	Tabulation	STUDYID, TDORDER	td.xpt	
TE	Trial Elements	Trial Design	One record per planned Element	Tabulation	STUDYID, ETCD	te.xpt	
TI	Trial Inclusion/Exclusion Criteria	Trial Design	One record per I/E criterion	Tabulation	STUDYID, IETESTCD	ti.xpt	
TS	Trial Summary	Trial Design	One record per trial summary parameter value	Tabulation	STUDYID, TSSEQ	ts.xpt	
TV	Trial Visits	Trial Design	One record per planned Visit per Arm	Tabulation	STUDYID, VISITNUM, ARMCD	tv.xpt	
DM	Demographics	Special Purpose	Special Purpose - One record per event per subject	Tabulation	STUDYID, USUBJID	dm.xpt	
EX	Exposure	Interventions	One record per constant dosing interval per subject	Tabulation	STUDYID, USUBJID, EXTRT, EXENDTC	ex.xpt	
AE	Adverse Events	Events	Events - One record per event per subject	Tabulation	STUDYID, USUBJID, AEDECOD, AESTDTC	ae.xpt	MedDRA dictionary version XX.1 was used to code all adverse events.
LB	Laboratory Test Results	Findings	Findings - One record per lab test per subject	Tabulation	STUDYID, USUBJID, LBCAT, LBTESTCD, VISITNUM	lb.xpt	

Variable-Level Metadata

Domain-level or dataset-level metadata is descriptive information about the SDTM domain itself. This metadata is used to define the domain or dataset variable content that is the ItemRef component of the define.xml file. It is also used within the variable definitions section or ItemDef of the define.xml file. This metadata is the workhorse of building the SDTM domains in Base SAS because it holds the definitional framework of each SDTM domain and variable. This metadata contains definitions such as your SAS label, SAS variable type, SAS variable length, any codelist name, and other associated metadata. Each row in this metadata represents a variable in an SDTM domain dataset.

The variable-level metadata is described in the following table.

Table 2.2: Variable-Level Metadata

Excel Column	Description
DOMAIN	This is the name of the SDTM domain and should match the NAME in the Table of Contents metadata file.
VARNUM	The order of the variable as it is to appear in the SDTM domain, as well as the order in which it appears in define.xml. You can get the general order of SDTM variables from the SDTM metadata spreadsheet sdtmv1_2_sdtmigv3_1_2_2009_03_27_final.xls. However, if you add variables to domains as the SDTM allows, then you need to insert those variables as best you can.
VARIABLE	The name of the SDTM variable required by CDISC.
TYPE	This should be `text`, `integer`, `float`, `datetime`, `date`, or `time`, as appropriate. Note that SAS datasets have types of only numeric or character, so you need to add this lower level of variable type specificity.
LENGTH	The length of the SDTM variable for TYPEs of `integer`, `float`, or `text` only. For SAS numeric variables, you can generally just set this to `8`. For character variables, you need to define what this length is. Note that CDISC does not dictate variable lengths. You should stop to deliberate about what your standard lengths should be. Also note that the FDA addressed the issue of having needlessly long text variables in the Study Data Technical Conformance Guide.
LABEL	The label of the SDTM variable required by CDISC.
SIGNIFICANTDIGITS	If type=`float`, then this should be populated with the integer representing the number of decimal places for the variable.

Excel Column	Description
ORIGIN	This text should describe where the variable comes from. ORIGIN types are under controlled terminology in define 2.0. Valid values are: `CRF`, `Derived`, `Assigned`, `Protocol`, `eDT`, and `Predecessor`. "CRF" may be followed by "Page" or "Pages" and the corresponding annotated CRF page numbers (for example, "CRF Page 1").
	If `Derived`, then a COMPUTATIONMETHODOID should be provided. `Predecessor` is typically used for ADaM data. Refer to Chapter 6 for further details.
COMMENTOID	This provides the OID to a row in the Comments spreadsheet.
DISPLAYFORMAT	This is the display format for numeric values.
COMPUTATIONMETHODOID	If the variable has a computational method, then this variable points to that computational method unique identifier (for example, AGECALCULATION).
CODELISTNAME	If the variable has controlled terminology, whether CDISC-supplied or not, this variable points to the unique identifier of that codelist. This CODELISTNAME value is also used to populate the CODELIST OID.
MANDATORY	This specifies whether an observation for the variable is mandatory to be present and not null. It can be populated with `Yes` or `No`. The authors suggest that SDTM variables that have a core definition of `Required` be set to `Yes`. Otherwise, this is set to `No`.
ROLE	In general, this is the role of the SDTM variable, as required by CDISC.
ROLECODELIST	Here we set this to ROLECODE everywhere as the unique identifier that points to the ROLECODE codelist for ROLE.
SASFIELDNAME	This can be left blank, and the %make_define2 macro will populate this element with same value as VARIABLE. Otherwise, if your SAS dataset variable name differs from the name that you wish to use in the SAS transport file in an FDA submission, then the SAS transport file variable name can be specified here.

To see the variable-level metadata that we use in this chapter, refer to "Appendix B.2 –Variable-Level Metadata." When the variable level metadata is put in the define.xml and rendered in a web browser, the section for the AE domain looks like this:

Adverse Events (AE) [Location: ./ae.xpt]

Variable	Label	Key	Type	Length	Controlled Terms or Format	Origin	Derivation/Comment
STUDYID	Study Identifier	1	text	15		Assigned	
DOMAIN	Domain Abbreviation		text	2		Derived	
USUBJID	Unique Subject Identifier	2	text	25		Assigned	
AESEQ	Sequence Number		integer	8		Derived	
AETERM	Reported Term for the Adverse Event		text	200		CRF Page 6	
AELLT	Lowest Level Term		text	200		Assigned	
AELLTCD	Lowest Level Term Code		integer	8		Assigned	
AEDECOD	Dictionary-Derived Term	3	text	200	AEDECOD	Assigned	
AEPTCD	Preferred Term Code		integer	8		Assigned	
AEHLT	High Level Term		text	200		Assigned	
AEHLTCD	High Level Term Code		integer	8		Assigned	
AEHLGT	High Level Group Term		text	200		Assigned	
AEHLGTCD	High Level Group Term Code		integer	8		Assigned	
AEBODSYS	Body System or Organ Class		text	200	AEBODSYS	Assigned	
AEBDSYCD	Body System or Organ Class Code		integer	8		Assigned	
AESOC	Primary System Organ Class		text	200		Assigned	
AESOCCD	Primary System Organ Class Code		integer	8		Assigned	
AESEV	Severity/Intensity		text	20	["MILD" = "Grade 1; 1", "MODERATE" = "Grade 2; 2", "SEVERE" = "Grade 3; 3"] <AESEV>	CRF Page 6	

Codelist Metadata

Codelists are the controlled terminology applied to your SDTM variables. An example of this controlled terminology is attached to the SEX variable in the DM domain. The SEX variable has the CDISC-defined controlled terminology set called "SEX," which includes M=Male, F=Female, and other codes. You will probably need to get your codelist data from multiple places. You can get the CDISC standard SDTM codelists by going to the CDISC website at http://www.cdisc.org/standards/foundational/terminology, which points you to the NCI Enterprise Vocabulary Services (EVS). EVS has the codelists available to you in Microsoft Excel format. You will also have other codelists that your data might use. These can be local to your company or to the trial. Finally, you might need to point to external codelists such as ones used for various medical dictionaries.

The codelist metadata is described in the following table.

Table 2.3: Codelist Metadata

Excel Column	Description
CODELISTNAME	The unique short name of the codelist. This is also used for the codelist OID in the define.xml file. For codelists that come from published CDISC-controlled terminology, you can find this name in the "CDISC submission value" column in blue highlighting. Otherwise, you can define this CODELISTNAME as you need to.
RANK	This is an optional rank order for the codelist item. If alphabetical order is sufficient, then RANK can be left blank. However, if the coded terms should be listed in a non-alphabetical order that makes more sense, then RANK should indicate that ordering.
CODEDVALUE	This is the value that you will find in the SDTM file in the result field, for example. For codelists that come from published CDISC-controlled terminology, you can find these values in the "CDISC submission value" column (not highlighted in blue). Otherwise, you can define these values as needed for your project.
TRANSLATED	The text "translation" of the CODEDVALUE, which is often the same as CODEDVALUE. For codelists that come from published CDISC-controlled terminology, you can find these translated values in the "CDISC synonym(s)" column. If the "CDISC synonym(s)" column is blank, you can copy the value of the "CDISC submission value" over for this value. If this is not for a CDISC-controlled term, then you can define these values as needed for your project.
TYPE	The data type of the codelist. This can be `text`, `float`, or `integer`.
CODELISTDICTIONARY	This is the name of the external codelist dictionary. It is populated for externally defined codelists only (for example, MedDRA).
CODELISTVERSION	This is populated for externally defined codelists only. It is the version number of CODELISTDICTIONARY.
ORDERNUMBER	Controls the order in which the codelist items appear in the define file.

Excel Column	Description
	Note that the following four variables are used in the Base SAS code in Chapter 3 to make value mapping from source to SDTM target easier. These variables are not used as part of making define.xml later in this chapter.
SOURCEDATASET	This variable points to the source dataset that contains the variable that will be mapped to the SDTM-controlled terminology. For codelists that do not have direct source data associations, this can be left blank.
SOURCEVARIABLE	This variable points to the source variable that will be mapped to the SDTM-controlled terminology. For codelists that do not have direct source data associations, this can be left blank.
SOURCEVALUE	This variable points to the source variable value that contains the actual value that will be mapped to a given SDTM-controlled terminology value. For codelists that do not have direct source data associations, this can be left blank.
SOURCETYPE	This should be set to `number` or `character`, depending on whether the SOURCEVARIABLE is numeric or character in SAS. For codelists that do not have direct source data associations, this can be left blank.

To see the codelist metadata that we use for this chapter, refer to "Appendix B.3 – Codelist Metadata." When the codelist metadata is put in the define.xml and rendered in a web browser, the section for the AESEV codelist looks like this:

AESEV [CodeList.AESEV]

Permitted Value (Code)	Display Value (Decode)
MILD	Grade 1; 1
MODERATE	Grade 2; 2
SEVERE	Grade 3; 3

Value-Level Metadata

Because the SDTM contains a lot of its data structured in a name/value pair and somewhat normalized data structure, there is sometimes a need to have more specificity about what a given record contains. This specificity can be stored in value-level metadata. You will notice that a lot of the same metadata fields are used in the variable-level metadata.

The value-level metadata is described in the following table.

Table 2.4: Value-Level Metadata

Excel Column	Description
DOMAIN	The domain for the source variable
VARIABLE	The name of the source variable for which value-level metadata is being provided.
WHERECLAUSEOID	Provides the OID for a row in the WHERE_CLAUSES spreadsheet.
VALUEVAR	The variable to be used by VALUENAME.
VALUENAME	Sets of records for which value-level metadata apply can be uniquely identified when the variable specified by VALUEVAR contains values that equal VALUENAME.
TYPE	This should be text, integer, float, datetime, date, or time, as appropriate. Note that SAS datasets have only numeric or character types, so you need to add this lower level of variable type specificity.
LENGTH	The length of the item for TYPEs of integer, float, or text only. For SAS numeric variables, you can generally just set this to 8. For character variables, you need to define what this length is. Note that CDISC does not require variable lengths. You should stop and deliberate about what your standard lengths should be.
LABEL	The label of the value-level item.
SIGNIFICANTDIGITS	If the type=float, then this should be populated with the integer representing the number of decimal places for the item.

Excel Column	Description
ORIGIN	This text should describe where the variable comes from. ORIGIN types are under controlled terminology in define 2.0. Valid values are: `CRF`, `Derived`, `Assigned`, `Protocol`, `eDT`, and `Predecessor`. "CRF" may be followed by "Page" or "Pages" and the corresponding annotated CRF page numbers (for example, "CRF Page 1").
	If `Derived`, then a COMPUTATIONMETHODOID or a COMMENTOID should be provided. `Predecessor` is another allowed value but is typically used for ADaM data. Refer to Chapter 6 for further details.
	If `Derived`, then a COMPUTATIONMETHODOID or a COMMENTOID should be provided. `Predecessor` is typically used for ADaM data. Refer to Chapter 6 for further details.
COMMENTOID	The unique ID for a comment that appears in the COMMENTS spreadsheet.
DISPLAYFORMAT	This is the display format for numeric items.
COMPUTATIONMETHODOID	If the value has a computational method, then this variable points to that computational method unique identifier (for example, AGECALCULATION) that appears in the COMPUTATION_METHOD spreadsheet.
CODELISTNAME	If the item has controlled terminology, whether CDISC-supplied or not, this variable points to the unique identifier of that code list. For codelists, we use CODELISTNAME to populate the CODELIST OID as well.
MANDATORY	This specifies whether the item is mandatory. It can be populated with `Yes` or `No`.
ROLE	In general, this is the role of the SDTM item as required by CDISC.
ROLECODELIST	We set this to ROLECODE for all records as the unique identifier that points to the ROLECODE codelist that defines the controlled terms for ROLE.

The first two fields, DOMAIN and VARIABLE, are used by the %make_define2 program to create a unique VALUELISTOID in the define file. This provides a standard naming convention and obviates the need to come up with one independently.

The WHERECLAUSEOID can be left blank for simple relationships based only on the variable specified by VALUEVAR being equal to the value specified by VALUENAME. In these situations, %make_define2 will create the where clause. This is the case for all of our example data except for one situation. Consider the glucose lab test. Regardless of whether glucose is measured by a blood chemistry panel or by a urinalysis, we should have LBTESTCD=GLUC. Other fields, such as LBCAT, can be used to distinguish between the two specimen types. Since the blood chemistry result is a numeric one while the urinalysis result is character based, with a discrete code list ("NEGATIVE" and "POSITIVE"), value-level metadata is definitely needed to set the two apart.

To see the value-level metadata that we use for this chapter, refer to "Appendix B.4 – Value-Level Metadata." When the value-level metadata is put in the define.xml and rendered in a web browser, you can see it in two places. If you look at the variable LBORRES in the LB variable metadata, you see it is a hyperlink:

Laboratory Tests (LB) [Location: ./lb.xpt]

Variable	Label	Key	Type	Length	Controlled Terms or Format	Origin	Derivation/Comment
STUDYID	Study Identifier	1	text	15		Assigned	
DOMAIN	Domain Abbreviation		text	2		Derived	
USUBJID	Unique Subject Identifier	2	text	25		Assigned	
LBSEQ	Sequence Number		integer	8		Derived	
LBTESTCD	Lab Test or Examination Short Name	4	text	8	LBTESTCD	Derived	
LBTEST	Lab Test or Examination Name		text	40	LBTEST	Derived	
LBCAT	Category for Lab Test	3	text	40	["CHEMISTRY" = "CHEMISTRY", "HEMATOLOGY" = "HEMATOLOGY", "URINALYSIS" = "URINALYSIS"] <LBCAT>	Derived	
LBORRES	Result or Finding in Original Units		text	200		eDT	
LBORRESU	Original Units		text	40	UNIT	eDT	

If you click on that hyperlink (./lb.xpt), it takes you to the value-level metadata table where you can see the value-level metadata for LBORRES:

Value Level Metadata - LB [LBORRES]

Variable	Where	Type	Length / Display Format	Controlled Terms or Format	Origin	Derivation/Comment
LBORRES	LBTESTCD EQ ALB (Albumin; Microalbumin)	float	4.2		eDT	
LBORRES	LBTESTCD EQ ALP (Alkaline Phosphatase)	float	3.		eDT	
LBORRES	LBTESTCD EQ ALT (Alanine Aminotransferase; SGPT)	float	3.		eDT	
LBORRES	LBTESTCD EQ AST (Aspartate Aminotransferase; SGOT)	float	3.		eDT	
LBORRES	LBTESTCD EQ BILDIR (Direct Bilirubin)	float	4.2		eDT	
LBORRES	LBTESTCD EQ BILI (Bilirubin; Total Bilirubin)	float	4.2		eDT	
LBORRES	LBTESTCD EQ GGT (Gamma Glutamyl Transferase)	float	3.		eDT	
LBORRES	LBCAT EQ CHEMISTRY (CHEMISTRY) AND LBTESTCD EQ GLUC (Glucose)	float	4.2		eDT	
LBORRES	LBCAT EQ URINALYSIS (URINALYSIS) AND LBTESTCD EQ GLUC (Glucose)	text	8	["NEGATIVE" = "Negative", "POSITIVE" = "Positive"] <URINGLUC>	eDT	
LBORRES	LBTESTCD EQ HCT (EVF; Erythrocyte Volume Fraction; Hematocrit; PCV; Packed Cell Volume)	float	3.		eDT	
LBORRES	LBTESTCD EQ HGB (FHGB; Free Hemoglobin; Hemoglobin)	float	5.2		eDT	
LBORRES	LBTESTCD EQ PROT (Protein)	float	4.2		eDT	

As discussed, most "values" are defined simply by LBTESTCD. But in the case of glucose, we also need to consider the value of LBCAT to distinguish between the blood chemistry and urinalysis results. The urinalysis result is the only character one, and it has a code list associated with it.

Where Clause Metadata

Where clause metadata is new to Define-XML version 2.0. The information needed to construct this type of metadata is provided in the following table.

Table 2.5: Where Clause Metadata

Excel Column	Description
WHERECLAUSEOID	An OID for the where clause.
SEQ	A counter to control the order in which the where clause is constructed.
SOFTHARD	If an actual data value fails the constraint, it is either rejected (a Hard constraint) or a warning is produced (a Soft constraint). In the context of Define-XML, this has no meaning. So all values may be set to Soft.
ITEMOID	The ITEMOID should uniquely identify the variable being compared. Usually [DOMAIN].[VARIABLE] should suffice.
COMPARATOR	The comparison operator. Valid values are LT, LE, GT, GE, EQ, NE, IN, and NOTIN.
VALUES	The actual data value used for the comparison.

Excel Column	Description
COMMENTOID	An optional reference to an OID in the Comments spreadsheet.

Although where clauses exist in the define.xml file for every value level metadata item, they only need to be specified in the metadata spreadsheet for records that cannot be uniquely identified by only one variable value. The %make_define2 macro will create them for these simple situations. For example, for the albumin value-level metadata, the macro creates a WHERECLAUSEOID equal to WC.LB.ORRES.ALB, and this where clause is designed to check for LBTESTCD EQ ALB. An example of the actual where clause XML is as follows:

```
<def:WhereClauseDef OID="WC.LB.LBORRES.ALB">
 <RangeCheck SoftHard="Soft" def:ItemOID="LB.LBTESTCD"
Comparator="EQ">
   <CheckValue>ALB</CheckValue>
 </RangeCheck>
</def:WhereClauseDef>
```

This is being shown to help prevent you from creating your own WHERECLAUSEOID that might conflict with the ones that %make_define2 automatically creates.

As mentioned earlier, the only where clauses that need to be constructed for our example data are those associated with the glucose lab test. We saw how this metadata gets represented in the define file in the previous section "*Value Level Metadata*." The following screen shot demonstrates how the glucose-related where clauses are captured in spreadsheet.

A	B	C	D	E	F	G
WHERECLAUSEOID	SEQ	SOFTHARD	ITEMOID	COMPARATOR	VALUES	COMMENTOID
WC.LB.LBTESTCD.GLUC.LBCAT.CHEMISTRY	1	Soft	LB.LBCAT	EQ	CHEMISTRY	
WC.LB.LBTESTCD.GLUC.LBCAT.CHEMISTRY	2	Soft	LB.LBTESTCD	EQ	GLUC	
WC.LB.LBTESTCD.GLUC.LBCAT.URINALYSIS	1	Soft	LB.LBCAT	EQ	URINALYSIS	
WC.LB.LBTESTCD.GLUC.LBCAT.URINALYSIS	2	Soft	LB.LBTESTCD	EQ	GLUC	

Computational Method Metadata

Although the CDISC SDTM can be described as containing primarily the collected data for a clinical trial, you can store some derived information in the SDTM. Total scores for a questionnaire, derived laboratory values, and even the AGE variable in DM can be considered derived content that is appropriate to store in the SDTM. The define.xml file gives you a place to store this derivation, and as such we have metadata to hold that information.

The computational method metadata is described in the following table.

Table 2.6: Computational Method Metadata

Excel Column	Description
COMPUTATIONMETHODOID	This variable points to that computational method unique identifier (for example, "AGECALCULATION").
LABEL	A descriptive label to associate with the computational method.
TYPE	Valid values are `Computation` and `Imputation`. A `Computation` uses an algorithm to derive a value. An `Imputation` is the process of replacing missing data with substitute values.
COMPUTATIONMETHOD	A text string that describes the computational method. This can be natural language text, pseudo code, or even executable code.

To see the computational algorithm metadata that we use for this chapter, refer to "Appendix B.5 –Computational Method Metadata." When the computational algorithm metadata is put in the define.xml and rendered in a web browser, the information about computational methods is in two places. The first is in the variable-level metadata section as is seen for AGE in the lower right corner of this snippet of the define file for the DM domain:

Demographics (DM) [Location: dm.xpt]

Variable	Label	Key	Type	Length	Controlled Terms or Format	Origin	Derivation/Comment
AGE	Age		integer	8		Derived	integer value of (BRTHDTC - RFSTDTC)/365.25

There is also a separate section of the define file where all computational methods are displayed. Navigating there, you can also find the age algorithm:

Computational Algorithms

Method	Type	Description
Algorithm to compute age	Computation	integer value of (BRTHDTC - RFSTDTC)/365.25

Comments Metadata

Any *ItemGroupDef* or *ItemDef* metadata element can have a comment associated with it, as we saw in many of the previous metadata sections. This is a change from version 1.0, which allowed specific comment attributes to each element. The change is a good one since it allows identical comments to be specified once, but referenced in multiple locations. The Comments metadata spreadsheet provides a common place to specify unique comments. The comments metadata is described in the following table.

Table 2.7: Comments Metadata

Excel Column	Description
COMMENTOID	The unique OID for the comment.
COMMENT	A comment itself, which is a free-text string

There is also a separate section of the define file where all unique comments are displayed. Navigating there, you can also find the two comments applied to our SDTM data:

Comments

CommentOID	Description
COM.ARMCD	Assigned based on Randomization Number.
com.ae	MedDRA dictionary version XX.1 was used to code all adverse events.

External Links

Typically, an SDTM define file has links to two documents: the annotated CRFs and a reviewer's guide. These files, and additional supplemental documents if desired, should now be specified in the External_Links spreadsheet. The External_Links metadata is described in the following table.

Table 2.8: External Links Metadata

Excel Column	Description
LeafID	The unique ID for the linked document
LeafRelPath	The relative path to the linked document, including the full filename
LeafPageRef	A specific named destination or page number within a PDF document.

Excel Column	Description
LeafPageRefType	Valid values are "PhysicalRef" and "NamedDestination." If LeafPageType=NamedDestination, then the NamedDestination must exist within the PDF document. If PhysicalRef, then a list of page numbers can be provided, separated by a space.
Title	A title for the document as you wish for it to appear in the define.xml file
SupplementalDoc	Valid values recognized by %make_define2 are Y, y, Yes, or 1. If any of these values are present, then the linked document will appear in the navigation pane.
AnnotatedCRF	Valid values recognized by %make_define2 are Y, y, Yes, or 1. If any of these values are present, then the linked annotated CRF document will appear in the navigation pane and will be designated as such in the XML. This designation will allow links from variables with an origin of CRF to target this document and the specified pages within the document.

For our SDTM example, we have two externally linked documents: the CRT reviewer's guide and the annotated CRF document. The following screenshot shows how these are represented in the External_Links spreadsheet.

LeafID	LeafRelPath	LeafPageRef	LeafPageRefType	Title	SupplementalDoc	AnnotatedCRF
blankcrf	blankcrf.pdf			Annotated Case Report Form		Y
CRTRG	reviewersguide.pdf			CRT Reviewer's Guide	Y	

Note that the LeafRelPath values specify only the filenames since these documents reside (or will reside) in the same directory as the define file. The corresponding XML in the define file is as follows:

```
<def:AnnotatedCRF>
  <def:DocumentRef leafID="blankcrf"/>
</def:AnnotatedCRF>
<def:leaf ID="blankcrf"
  xlink:href="blankcrf.pdf">
  <def:title>Annotated Case Report Form </def:title>
</def:leaf>
<def:SupplementalDoc>
  <def:DocumentRef leafID="CRTRG"/>
</def:SupplementalDoc>
<def:leaf ID="CRTRG"
  xlink:href="reviewersguide.pdf">
  <def:title>CRT Reviewer's Guide </def:title>
</def:leaf>
```

Building Define.xml

A significant benefit to using metadata files to drive your SDTM domain creation is that you have the essential ingredients to build define.xml. Each of the spreadsheets introduced earlier in this chapter represent separate parts of the metadata file:

- Table of Contents metadata file
- Variable-level metadata file
- Codelist metadata file
- Value-level metadata file
- Where clause metadata file
- Computational method metadata file
- Comments metadata file
- External links metadata file

Define File Header Metadata

Now you just need to define one more set of metadata, and you can create the define.xml file. This last remaining piece is the header information for define.xml. That metadata file is described in the following table.

Table 2.9: Metadata File for Define.xml

Excel Column	Description
FILEOID	A unique identifier for the define.xml file.
STUDYOID	A unique identifier for the study.
STUDYNAME	A short name of the study.
STUDYDESCRIPTION	A long text description of the study.
PROTOCOLNAME	A short name of the study. Can be equal to STUDYNAME.
STANDARD	The CDISC standard (in this case, SDTM. In Chapter 6, it would be ADaM).
VERSION	The version number of the standard.
SCHEMALOCATION	The location of the XML schema file that defines the structure of define.xml. Currently not used by %make_define2.
STYLESHEET	The name of the XSL style sheet that will be used to render the define.xml file in your web browser. Note that in this book, the style sheet used for the SDTM is the one that was released with the define.xml version 1.0.

To see the define header metadata that we use for this chapter, refer to "Appendix B.6 – Define Header Metadata." This metadata as defined is essentially metadata about define.xml itself, so it does not actually render visually in the define file when opened by a web browser.

Define File Creation SAS Program

Refer to the authors' SAS web pages (http://support.sas.com/publishing/authors/index.html for the %make_define2 SAS macro code that is used to generate define.xml for this SDTM data. The call to make the define.xml file looks like this:

```
%make_define2(path=C:\SDTM_metadata, metadata=SDTM_METADATA.xlsx)
```

In this call, the path macro parameter points to the folder where the metadata spreadsheet and the define file are finally stored. The metadata parameter points to the name of the SDTM metadata spreadsheet that contains tabs for all of the metadata defined above. The %make_define2 SAS macro assumes that your XSL style sheet that is defined in the define file header metadata field STYLESHEET already exists in the folder that is defined by the path macro parameter.

Chapter Summary

The best approach to implementing the SDTM in Base SAS is through the heavy use of metadata to drive your conversion process.

Build a metadata repository to hold your SDTM metadata, both to make conversions easier and to make building your define.xml file possible.

Leverage your metadata files to build your define.xml file with a program like make_define2.sas. You can customize this program as you see fit to create more metadata-driven parameters as you need to.

Chapter 3: Implementing the CDISC SDTM with Base SAS

This chapter provides an illustrated example of how you can implement the CDISC SDTM using Base SAS programming and the SDTM metadata from the previous chapter. Various SAS macros are used to help you easily create SDTM datasets. Several domain conversion program examples are presented.

More Information

    Appendix A - Source Data Programs

    Appendix B - SDTM Metadata

See Also

    Study Data Tabulation Model Implementation Guide version 3.2

    Define-XML version 2.0

Base SAS Macros and Tools for SDTM Conversions

Many of the tasks involved in transforming the clinical source data into the CDISC SDTM are repetitive in nature. The SAS Macro Language is an excellent tool for automating these repetitive tasks. This section of the chapter provides several SAS macros and a SAS format library program that will be used in the individual SDTM transformations that follow in the next section. The SAS code detailed in this section relies heavily on using the metadata files defined in the previous chapter. If you have additional repetitive SDTM tasks, then you can create additional SAS macros, in addition to the ones provided here.

A basic flow of the SDTM creation process would follow the Extract-Transform-Load (ETL) paradigm of data warehousing. You first need to get your data, and then manipulate it to meet your needs. Finally, you store the data where you need it. The following table summarizes the SAS macros that help with the data transformation process, in rough chronological order.

Table 3.1: Base SAS Programs for SDTM Conversions

SAS Program or Macro	Purpose
make_codelist_formats.sas	This program takes the controlled terminology and codelist information from the SDTM metadata in Chapter 2 and creates a permanent SAS format library from it. That format library is used later in the data transformation process when source data values are mapped to target SDTM data values using PUT statements in a DATA step.
make_empty_dataset.sas	This macro creates an empty dataset shell based on the SDTM metadata from Chapter 2. The macro uses the metadata from Chapter 2 to define which variables belong in a given SDTM domain, which is the beginning of the transformation step. This macro also creates a macro variable that can be used at the load step to keep only the variables necessary per metadata requirements.
make_dtc_date.sas	This macro creates an ISO8601 date string from SAS date or datetime component variables. It is used during the DATA step transformation process.
make_sdtm_dy.sas	This macro creates an SDTM study day (*DY) variable when given two SDTM **DTC dates. It is used during the DATA step transformation process.
make_sort_order.sas	This macro creates a macro variable that can be used in the sort process for the final data load. The variable is based on the required sort order as specified in the metadata definition from Chapter 2. It is generally useful at the data loading stage after the data transformation has taken place.

Creating an SDTM Codelist SAS Format Catalog

Controlled terminology is a critical component of the SDTM. SDTM-controlled terminology is used within the SDTM datasets, and it is also included in the define.xml file. In this chapter, we need that controlled terminology to be applied to the source datasets in order to map the data records properly to the SDTM records. The codelist metadata file listed in the previous chapter is used here to create a permanent SAS format library that can be used in the SDTM conversions to follow. Here is the SAS macro program that will create the SAS format library for you.

```
*-------------------------------------------------------------------*;
* make_codelist_formats.sas creates a permanent SAS format library
* stored to the libref LIBRARY from the codelist metadata file
* CODELISTS.xls.  The permanent format library that is created
* contains formats that are named like this:
*    CODELISTNAME_SOURCEDATASET_SOURCEVARIABLE
* where CODELISTNAME is the name of the SDTM codelist,
* SOURCEDATASET is the name of the source SAS dataset and
* SOURCEVARIABLE is the name of the source SAS variable.
*-------------------------------------------------------------------*;
proc import
    datafile="SDTM_METADATA.xls"
    out=formatdata
    dbms=excelcs
    replace;
    sheet="CODELISTS";
run;


** make a proc format control dataset out of the SDTM metadata;
data source.formatdata;
    set formatdata(drop=type);

        where sourcedataset ne "" and sourcevalue ne "";

        keep fmtname start end label type;
        length fmtname $ 32 start end $ 16 label $ 200 type $ 1;

        fmtname = compress(codelistname || "_" || sourcedataset
                    || "_" || sourcevariable);
        start = left(sourcevalue);
        end = left(sourcevalue);
        label = left(codedvalue);
        if upcase(sourcetype) = "NUMBER" then
            type = "N";
        else if upcase(sourcetype) = "CHARACTER" then
            type = "C";
run;

** create a SAS format library to be used in SDTM conversions;
proc format
    library=library
    cntlin=source.formatdata
    fmtlib;
run;
```

This program does assume that you have defined the SAS libref called library to store these permanent SAS formats. The name library is used so that SAS can automatically see the SAS formats without having to specify FMTSEARCH explicitly in the OPTIONS statement.

After you run the program, if you look at the resulting SAS format entry from the source TRT variable in source dataset DEMOGRAPHIC, you will see this:

FORMAT NAME: ARM_DEMOGRAPHIC_TRT LENGTH: 19 MIN LENGTH: 1 MAX LENGTH: 40 DEFAULT LENGTH 19 FUZZ: STD		
START	END	LABEL (VER. 9.4 22 NOV 2015:07:11:36)
0 1	0 1	Placebo Analgezia HCL 30 mg

This gives us the ARM_DEMOGRAPHIC_TRT format that can be used to map the DEMOGRAPHIC TRT variable into the ARM variable in the SDTM DM dataset.

Creating an Empty SDTM Domain Dataset

Because we went to great lengths to define the SDTM domain-level metadata, it makes sense that we leverage that in our program code. The following SAS macro reads that domain-level metadata spreadsheet and creates an empty SAS dataset that you can populate. Here is that SAS macro program and details about how it works.

```
*------------------------------------------------------------*;
* make_empty_dataset.sas creates a zero record dataset based on a dataset
* metadata spreadsheet. The dataset created is calledEMPTY_** where "**"
* is the name of the dataset. This macro also creates a global macro
* variable called **KEEPSTRING that holds the dataset variables desired
* and listed in the order they should appear. [The variable order is
* dictated by VARNUM in the metadata spreadsheet.]
*
* MACRO PARAMETERS:
* metadatafile = the MS Excel file containing VARIABLE_METADATA
* dataset = the dataset or domain name you want to extract
*------------------------------------------------------------*;
%macro make_empty_dataset(metadatafile=,dataset=);

    proc import ❶
        datafile="&metadatafile"
        out=_temp
        dbms=excelcs
        replace;
        sheet="VARIABLE_METADATA";
    run;

    ** sort the dataset by expected specified variable order; ❷
    proc sort
      data=_temp;
        where domain = "&dataset";
        by varnum;
    run;

    ** create keepstring macro variable and load metadata
    ** information into macro variables;  ❸
    %global &dataset.KEEPSTRING;
    data _null_;
```

```
      set _temp nobs=nobs end=eof;

        if _n_=1 then
          call symput("vars", compress(put(nobs,3.)));

        call symputx('var'    || compress(put(_n_, 3.)),variable);
        call symputx('label'  || compress(put(_n_, 3.)), label);
        call symputx('length' || compress(put(_n_, 3.)),
                                        put(length, 3.));

        ** valid ODM types include TEXT, INTEGER, FLOAT, DATETIME,
        ** DATE, TIME and map to SAS numeric or character;
        if upcase(type) in ("INTEGER", "FLOAT") then
          call symputx('type' || compress(put(_n_, 3.)), "");
        else if upcase(type) in ("TEXT", "DATE", "DATETIME",
                                 "TIME") then
          call symputx('type' || compress(put(_n_, 3.)), "$");
        else
          put "ERR" "OR: not using a valid ODM type.  " type=;

        ** create **KEEPSTRING macro variable;       ❹
        length keepstring $ 32767;
        retain keepstring;
        keepstring = compress(keepstring) || "|" ||left(variable);
        if eof then
          call symputx(upcase(compress("&dataset"||'KEEPSTRING')),
                       left(trim(translate(keepstring," ","|"))));
      run;


    ** create a 0-observation template data set used for assigning
    ** variable attributes to the actual data sets;    ❺
    data EMPTY_&dataset;
        %do i=1 %to &vars;
            attrib &&var&i label="&&label&i"
                   length=&&type&i.&&length&i...
            ;
            %if &&type&i=$ %then
              retain &&var&i '';
            %else
              retain &&var&i .;
            ;
        %end;
        if 0;
    run;
%mend make_empty_dataset;
```

When the %make_empty_dataset macro is executed, the EMPTY_** dataset is created, and the **KEEPSTRING global macro variable is defined.

❶   The Microsoft Excel metadata file used here is the domain-level metadata file described in the previous chapter.

❷   The domain variables are sorted by the variable order specified in the VARNUM variable. This
sort order is also used to order the variables in the **KEEPSTRING global macro variable.

❸   This DATA step loads the domain metadata that we need into VAR*, LABEL*, LENGTH*, and TYPE* macro parameters for each variable in the domain to be used in the next step.

❹   This section is responsible for defining the **KEEPSTRING global macro variable, which will be used in the actual domain creation code later.

❺   This DATA step defines the SAS work EMPTY_** dataset, which is the shell of the domain that we will populate later.

Suppose you submit the macro like this for the DM domain:

```
%make_empty_dataset(metadatafile=SDTM_METADATA.xls,dataset=DM)

proc contents
  data=work.empty_dm;
run;
```

You get a subsequent empty SAS dataset in SASWORK that looks like this when you run a PROC CONTENTS on it:

```
The CONTENTS Procedure


Data Set Name        WORK.EMPTY_DM          Observations              0
Member Type          DATA                   Variables                23
Engine               V9                     Indexes                   0
Created              11/27/2015 10:42:47    Observation Length      392
Last Modified        11/27/2015 10:42:47    Deleted Observations      0
Protection                                  Compressed               NO
Data Set Type                               Sorted                   NO
Label
Data Representation  WINDOWS_64
Encoding             wlatin1  Western (Windows)


                                        Engine/Host Dependent
      Information

      Data Set Page Size          65536
      Number of Data Set Pages    1
      First Data Page             1
      Max Obs per Page            167
      Obs in First Data Page      0
      Number of Data Set Repairs  0
      ExtendObsCounter            YES
      Filename                    empty_dm.sas7bdat
      Release Created             9.0401M2
      Host Created                X64_7PRO
```

```
                Alphabetic List of Variables and Attributes

    #      Variable     Type     Len      Label

    22     ACTARM       Char      40      Description of Actual Arm
    21     ACTARMCD     Char       8      Actual Arm Code
    15     AGE          Num        8      Age
    16     AGEU         Char      10      Age Units
    20     ARM          Char      40      Description of Planned Arm
    19     ARMCD        Char       8      Planned Arm Code
    14     BRTHDTC      Char      16      Date/Time of Birth
    23     COUNTRY      Char       3      Country
     2     DOMAIN       Char       2      Domain Abbreviation
    11     DTHDTC       Char      16      Date/Time of Death
    12     DTHFL        Char       2      Subject Death Flag
    18     RACE         Char      80      Race
     6     RFENDTC      Char      16      Subject Reference End Date/Time
     9     RFICDTC      Char      16      Date/Time of Informed Consent
    10     RFPENDTC     Char      16      Date/Time of End of Participation
     5     RFSTDTC      Char      16      Subject Reference Start Date/Time
     8     RFXENDTC     Char      16      Date/Time of Last Study Treatment
     7     RFXSTDTC     Char      16      Date/Time of First Study Treatment
    17     SEX          Char       2      Sex
    13     SITEID       Char       7      Study Site Identifier
     1     STUDYID      Char      15      Study Identifier
     4     SUBJID       Char       7      Subject Identifier for the Study
     3     USUBJID      Char      25      Unique Subject Identifier
```

Creating an SDTM --DTC Date Variable

Dates and datetimes in the SDTM are all presented as ISO8601 character text strings. These date strings can be created from partial date information. Here is a SAS macro program that creates an SDTM --DTC date variable for you within a DATA step when given the numeric parts of a SAS date or datetime.

```
*------------------------------------------------------------------*;
* make_dtc_date.sas is a SAS macro that creates a SDTM --DTC date
* within a SAS datastep when provided the pieces of the date in
* separate SAS variables.
*
* NOTE: This macro must have SAS OPTIONS MISSING = ' ' set before
* it is called to handle missing date parts properly.
*
* MACRO PARAMETERS:
* dtcdate = SDTM --DTC date variable desired
* year = year variable
* month = month variable
* day = day variable
* hour = hour variable
* minute = minute variable
* second = second variable
*------------------------------------------------------------------*;
   %macro make_dtc_date(dtcdate=, year=., month=., day=.,
                        hour=., minute=., second=.);
```

```
        ** in a series of if-then-else statements, determine where the
        ** smallest unit of date and time is present and then construct a DTC
        ** date based on the non-missing date variables.;

        if (&second ne .) then
           &dtcdate = put(&year,z4.) || "-" || put(&month,z2.) || "-"
                          || put(&day,z2.) || "T" || put(&hour,z2.) || ":"
                          || put(&minute,z2.) || ":" || put(&second,z2.);
        else if (&minute ne .) then
           &dtcdate = put(&year,z4.) || "-" || put(&month,z2.) || "-"
                          || put(&day,z2.) || "T" || put(&hour,z2.) || ":"
                          || put(&minute,z2.);
        else if (&hour ne .) then
           &dtcdate = put(&year,z4.) || "-" || put(&month,z2.) || "-"
                          || put(&day,z2.) || "T" || put(&hour,z2.);
        else if (&day ne .) then
           &dtcdate = put(&year,z4.) || "-" || put(&month,z2.) || "-"
                          || put(&day,z2.);
        else if (&month ne .) then
           &dtcdate = put(&year,z4.) || "-" || put(&month,z2.);
        else if (&year ne .) then
           &dtcdate = put(&year,z4.);
        else if (&year = .) then
           &dtcdate = "";

        ** remove duplicate blanks and replace space with a dash;
           &dtcdate = translate(trim(compbl(&dtcdate)),'-',' ');
   %mend make_dtc_date;
```

A sample call of this SAS macro for the EX domain might look like this:

```
   data ex;

      %make_dtc_date(dtcdate=exstdtc, year=startyy,
                     month=startmm, day=startdd)
      %make_dtc_date(dtcdate=exendtc, year=endyy,
                     month=endmm, day=enddd,
                     hour=endhh, minute=endmi, second=endss)
   run;
```

That macro call to %make_sdtm_date would create the variable EXSTDTC in the resulting EX dataset in the YYYY-MM-DD format and EXENDTC in the YYYY-MM-DDTHH:MM:SS format.

Creating an SDTM Study Day Variable

Throughout the SDTM, you will find that you need to create "study day" or SDTM --DY variables. Because the mechanics of doing this are the same everywhere, the task lends itself to using standardized SAS macro code. Here is the SAS macro program that creates an SDTM --DY variable for you.

```
   *-------------------------------------------------------------------*;
   * make_sdtm_dy.sas is a SAS macro that takes two SDTM --DTC dates
   * and calculates a SDTM study day (--DY) variable.  It must be used
   * in a datastep that has both the REFDATE and DATE variables
   * specified in the macro parameters below.
   * MACRO PARAMETERS:
   * refdate = --DTC baseline date to calculate the --DY from.
```

```
*              This should be DM.RFSTDTC for SDTM --DY variables.
* date = --DTC date to calculate the --DY to.  The variable
*            associated with the --DY variable.
*-------------------------------------------------------------*;

%macro make_sdtm_dy(refdate=RFSTDTC,date=);

    if length(&date) >= 10 and length(&refdate) >= 10 then
      do;
        if input(substr(%substr("&date",2,%length(&date)-
          3)dtc,1,10),yymmdd10.) >=
          input(substr(%substr("&refdate",2,%length(&refdate)-
          3)dtc,1,10),yymmdd10.) then
          %upcase(%substr("&date",2,%length(&date)-3))DY =
          input(substr(%substr("&date",2,%length(&date)-
          3)dtc,1,10),yymmdd10.) -
          input(substr(%substr("&refdate",2,%length(&refdate)-
          3)dtc,1,10),yymmdd10.) + 1;
        else
          %upcase(%substr("&date",2,%length(&date)-3))DY =
          input(substr(%substr("&date",2,%length(&date)-
          3)dtc,1,10),yymmdd10.) -
          input(substr(%substr("&refdate",2,%length(&refdate)-
          3)dtc,1,10),yymmdd10.);
      end;

%mend make_sdtm_dy;
```

A sample call of this SAS macro for the LB domain might look like this:

```
data lb;
    merge lb(in=inlb) target.dm(keep=usubjid rfstdtc);
          by usubjid;
              if inlb;
          %make_sdtm_dy(date=lbdtc)
run;
```

That macro call to %make_sdtm_dy would create the variable LBDY in the resulting LB dataset.

Sorting the Final SDTM Domain Dataset

In the Table of Contents section of the define.xml file is a field that defines how a domain is sorted. The following SAS macro takes the metadata for that sort order and creates a global SAS macro variable called **SORTSTRING, where **is the domain of interest. Keep in mind that this sort sequence is also what is likely used to define the --SEQ variable in the SDTM dataset. Here is that SAS macro program and details about how it works.

```
*-------------------------------------------------------------*;
* make_sort_order.sas creates a global macro variable called
* **SORTSTRING where ** is the name of the dataset that contains
* the KEYSEQUENCE metadata specified sort order for a given
* dataset.                                                  *;
```

```
* MACRO PARAMETERS:
* metadatafile = the file containing the dataset metadata
* dataset = the dataset or domain name
*------------------------------------------------------------*;
%macro make_sort_order(metadatafile=,dataset=);

    proc import                                     ❶
        datafile="&metadatafile"
        out=_temp
        dbms=excelcs
        replace;
        sheet="VARIABLE_METADATA";
    run;

    proc sort
        data=_temp;
        where keysequence ne . and domain="&dataset";
        by keysequence;
    run;

    ** create **SORTSTRING macro variable;      ❷
    %global &dataset.SORTSTRING;
    data _null_;
        set _temp end=eof;
        length domainkeys $ 200;
         retain domainkeys '';

         domainkeys = trim(domainkeys) || ' ' ||
trim(put(variable,8.));

         if eof then
            call symputx(compress("&dataset" || "SORTSTRING"),
domainkeys);
    run;

%mend make_sort_order;
```

When the %make_sort_order macro is executed, the **SORTSTRING global macro variable is created.

❶ The Excel metadata here is the VARIABLE_METADATA metadata tab described in the previous chapter.

❷ The **SORTSTRING variable is created in this step, which is used in the final dataset sorting when the actual domain is created. This is based on the order specified in the KEYSEQUENCE column in the VARIABLE_METADATA tab of the metadata spreadsheet.

Building SDTM Datasets

Now that you have your metadata store and your SAS macro library at hand, you can build the SDTM domains in Base SAS. In this section, we look at creating six different types of SDTM data. First, we create the special-purpose DM domain because it is needed to create the study day (--DY) variables in the domain datasets. Then we create a supplemental qualifier, findings, events, and interventions, and finally some trial design model datasets.

Building the Special-Purpose DM and SUPPDM Domains

The following SAS code builds the DM and SUPPDM datasets from our source clinical trial data found in Appendix A, "Source Data Programs." DM and SUPPDM represent the demographics domain in the SDTM. The following program assumes that the source datasets can be found under the libref `source` and that the permanent SAS formats can be found under the libref `library`.

```
*-------------------------------------------------------------*;
* DM.sas creates the SDTM DM and SUPPDM datasets and saves them
* as permanent SAS datasets to the target libref.
*-------------------------------------------------------------*;

**** CREATE EMPTY DM DATASET CALLED EMPTY_DM;                     ❶
%make_empty_dataset(metadatafile=SDTM_METADATA.xls,dataset=DM)

**** GET FIRST AND LAST DOSE DATE FOR RFSTDTC AND RFENDTC;        ❷
proc sort
  data=source.dosing(keep=subject startdt enddt)
  out=dosing;
    by subject startdt;
run;

**** FIRSTDOSE=FIRST DOSING AND LASTDOSE=LAST DOSING;
data dosing;
  set dosing;
    by subject;

    retain firstdose lastdose;

    if first.subject then
      do;
        firstdose = .;
        lastdose = .;
      end;
```

```
      firstdose = min(firstdose,startdt,enddt);
      lastdose = max(lastdose,startdt,enddt);

      if last.subject;
  run;

  **** GET DEMOGRAPHICS DATA;
  proc sort
    data=source.demographic
    out=demographic;
      by subject;
  run;

  **** MERGE DEMOGRAPHICS AND FIRST DOSE DATE;
  data demog_dose;
    merge demographic
          dosing;
      by subject;
  run;

  **** DERIVE THE MAJORITY OF SDTM DM VARIABLES;      ❸
  options missing = ' ';
  data dm;
    set EMPTY_DM
        demog_dose(rename=(race=_race));

      studyid = 'XYZ123';
      domain = 'DM';
      usubjid = left(uniqueid);                       ❹
      subjid = put(subject,3.);
      rfstdtc = put(firstdose,yymmdd10.);
      rfendtc = put(lastdose,yymmdd10.);
      rfxstdtc = put(firstdose,yymmdd10.);
      rfxendtc = put(lastdose,yymmdd10.);
      rficdtc = put(icdate,yymmdd10.);
      rfpendtc = put(lastdoc,yymmdd10.);
      dthfl = 'N';
      siteid = substr(subjid,1,1) || "00";
      brthdtc = put(dob,yymmdd10.);
      age = floor ((intck('month',dob,firstdose) -
            (day(firstdose) < day(dob))) / 12);
      if age ne . then
        ageu = 'YEARS';
      sex = put(gender,sex_demographic_gender.);      ❺
      race = put(_race,race_demographic_race.);
      armcd = put(trt,armcd_demographic_trt.);
      arm = put(trt,arm_demographic_trt.);
      actarmcd = put(trt,armcd_demographic_trt.);
      actarm = put(trt,arm_demographic_trt.);
      country = "USA";
  run;

  **** DEFINE SUPPDM FOR OTHER RACE;
  **** CREATE EMPTY SUPPDM DATASET CALLED EMPTY_SUPPDM;      ❻
  %make_empty_dataset(metadatafile=SDTM_METADATA.xls,dataset=SUPPDM)
```

```
data suppdm;
  set EMPTY_SUPPDM
      dm;

    keep &SUPPDMKEEPSTRING;                                          ❼

    **** OUTPUT OTHER RACE AS A SUPPDM VALUE;                        ❽
    if orace ne '' then
      do;
        rd1omain = 'DM';
        qnam = 'RACEOTH';
        qlabel = 'Race, Other';
        qval = left(orace);
        qorig = 'CRF Page 1';
        output;
      end;

    **** OUTPUT RANDOMIZATION DATE AS SUPPDM VALUE;
    if randdt ne . then
      do;
        rdomain = 'DM';
        qnam = 'RANDDTC';
        qlabel = 'Randomization Date';
        qval = left(put(randdt,yymmdd10.));
        qorig = 'CRF Page 1';
        output;
      end;
run;


**** SORT DM ACCORDING TO METADATA AND SAVE PERMANENT DATASET;      ❾
%make_sort_order(metadatafile=SDTM_METADATA.xls,dataset=DM)

proc sort
  data=dm(keep = &DMKEEPSTRING)
  out=target.dm;
    by &DMSORTSTRING;
run;


**** SORT SUPPDM ACCORDING TO METADATA AND SAVE PERMANENT DATASET;
%make_sort_order(metadatafile=SDTM_METADATA.xls,dataset=SUPPDM)

proc sort
  data=suppdm
  out=target.suppdm;
    by &SUPPDMSORTSTRING;
run;
```

When the DM.sas program has been run, the DM and SUPPDM SDTM domain datasets are saved to the target libref.

❶   The first step in this program creates the empty SDTM DM dataset called EMPTY_DM, based on the domain metadata spreadsheet.

❷   The SDTM DM variables RFSTDTC and RFENDTC are defined here as the first day of study medication dosing and the last day of dosing, respectively. In this example, there are no dosing times, just dates. Keep in mind that the use of the variables RFSTDTC and RFENDTC as defined here, based on dosing dates, is just one way of doing this. See the "SDTM Implementation Guide" for additional details about how you can define RFSTDTC and RFENDTC.

❸ This is where the empty dataset EMPTY_DM with the defining variable attributes is set with the source data, and where the bulk of the SDTM DM variables are defined. Note the RENAME clause in the SET statement, which is required when you have source variables named the same as target variables with conflicting variable attributes. Also note that EMPTY_DM is set first so that the variable attributes are maintained for the DATA step regardless of the contents of demog_dose.

❹ In these examples, you see that there is a source variable called UNIQUEID in the source datasets, which is supposed to uniquely identify a patient within and across studies. In most clinical databases, a truly unique patient identifier UNIQUEID does not exist, and, in practice, many trials are submitted where USUBJID is set equal to SUBJID.

❺ For SEX, RACE, ARM, and ARMCD variables, the associated formats were created in the make_codelist_formats.sas program in the prior section.

❻ The %make_empty_dataset macro is called again. This time it defines the dataset structure for the SUPPDM dataset.

❼ The SUPPDMKEEPSTRING macro variable created by %make_empty_dataset lists only the SDTM variables that we want in SUPPDM.

❽ The supplemental qualifiers created here in SUPPDM are for other race and randomization date. This chunk of code creates those qualifier records. However, you typically have to define IDVAR and IDVARVAL explicitly. In this case, they just default to null. IDVAR defines the SDTM parent domain variable. IDVARVAL defines the value of that variable that can be used to join or merge these supplemental qualifier variables back onto the parent domain. Often, IDVAR=--SEQ and IDVARVAL is the sequence number because --SEQ tends to be the most exacting identifier that you can have in the SDTM because it identifies a single record. You should define IDVAR to be the highest-level identifier that you can in order to be clear about the qualifying relationship to the parent domain.

❾ At this point in the program, we call the %make_sort_order macro twice to get the DMSORTSTRING and SUPPDMSORTSTRING macro variables defined. Then we use that to sort and save our final DM and SUPPDM domain datasets.

Building the LB Findings Domain

The following SAS code builds the LB laboratory data dataset from source clinical trial data found in "Appendix A - Source Data Programs." The following program assumes that the source datasets can be found under the libref `source` and that the permanent SAS formats can be found under the libref `library`.

```
*----------------------------------------------------------------*;
* LB.sas creates the SDTM LB dataset and saves it
* as a permanent SAS datasets to the target libref.
*----------------------------------------------------------------*;

**** CREATE EMPTY DM DATASET CALLED EMPTY_LB;
%make_empty_dataset(metadatafile=SDTM_METADATA.xls,dataset=LB)     ❶
```

```
**** DERIVE THE MAJORITY OF SDTM LB VARIABLES;
options missing = ' ';
data lb;
  set EMPTY_LB
      source.labs;

    studyid = 'XYZ123';
    domain = 'LB';
    usubjid = left(uniqueid);
    lbcat = put(labcat,$lbcat_labs_labcat.);              ❷
    lbtest = put(labtest,$lbtest_labs_labtest.);
    lbtestcd = put(labtest,$lbtestcd_labs_labtest.);
    lborres = left(put(nresult,best.));
    lborresu = left(colunits);
    lbornrlo = left(put(lownorm,best.));
    lbornrhi = left(put(highnorm,best.));

    **** create standardized results;                     ❸
    lbstresc = lborres;
    lbstresn = nresult;
    lbstresu = lborresu;
    lbstnrlo = lownorm;
    lbstnrhi = highnorm;

    **** urine glucose adjustment;
    if lbtest = 'Glucose' and lbcat = 'URINALYSIS' then
      do;
         lborres = left(put(nresult,uringluc_labs_labtest.));
         lbornrlo = left(put(lownorm,uringluc_labs_labtest.));
         lbornrhi = left(put(highnorm,uringluc_labs_labtest.));
         lbstresc = lborres;
         lbstresn = .;
         lbstnrlo = .;
         lbstnrhi = .;
      end;

    if lbtestcd = 'GLUC' and lbcat = 'URINALYSIS' and
      lborres = 'POSITIVE' then
      lbnrind = 'HIGH';
    else if lbtestcd = 'GLUC' and lbcat = 'URINALYSIS' and
            lborres = 'NEGATIVE' then
             lbnrind = 'NORMAL';
    else if lbstnrlo ne . and lbstresn ne . and
      round(lbstresn,.0000001) < round(lbstnrlo,.0000001) then
      lbnrind = 'LOW';
    else if lbstnrhi ne . and lbstresn ne . and
      round(lbstresn,.0000001) > round(lbstnrhi,.0000001) then
      lbnrind = 'HIGH';
    else if lbstnrhi ne . and lbstresn ne . then
      lbnrind = 'NORMAL';

    visitnum = month;
    visit = put(month,visit_labs_month.);

    if visit = 'Baseline' then                            ❹
      lbblfl = 'Y';
    else
```

```
      lbblfl = ' ';

    if visitnum < 0 then
      epoch = 'SCREENING';
    else
      epoch = 'TREATMENT';

    lbdtc = put(labdate,yymmdd10.);
run;


proc sort
  data=lb;
    by usubjid;
run;

**** CREATE SDTM STUDYDAY VARIABLES;            ❺
data lb;
  merge lb(in=inlb) target.dm(keep=usubjid rfstdtc);
    by usubjid;

    if inlb;

    %make_sdtm_dy(date=lbdtc)
run;


**** CREATE SEQ VARIABLE;
proc sort
  data=lb;
    by studyid usubjid lbcat lbtestcd visitnum;
run;

data lb;
  retain &LBKEEPSTRING;                         ❻
  set lb(drop=lbseq);
    by studyid usubjid lbcat lbtestcd visitnum;

   if not (first.visitnum and last.visitnum) then
    put "WARN" "ING: key variables do not define an unique record. "
          usubjid=;

    retain lbseq 0;
    lbseq = lbseq + 1;

    label lbseq = "Sequence Number";
run;


**** SORT LB ACCORDING TO METADATA AND SAVE PERMANENT DATASET;   ❼
%make_sort_order(metadatafile=SDTM_METADATA.xls,dataset=LB)
```

```
proc sort
  data=lb(keep = &LBKEEPSTRING)
  out=target.lb;
    by &LBSORTSTRING;
run;
```

When the LB.sas program has been run, the LB SDTM domain datasets are saved to the target libref.

❶ The first step in this program creates the empty SDTM LB dataset called EMPTY_LB based on the domain metadata spreadsheet. That is then set with the source `labs` dataset, and variable population begins.

❷ Note here that these are character SAS formats that are created by the make_codelist_formats.sas program and that hold the controlled terminology for LBCAT, LBTEST, and LBTESTCD.

❸ This is a very simplistic example where the collected results are mapped directly over to the standardized results. This is not very likely. Hopefully you will at least be given source data, perhaps in the CDISC LAB format, that has both standardized and collected results already provided for you. If not, mapping lab data from collected data into standardized results can be a labor-intensive task.

❹ LBBLFL is your baseline lab flag. Once again, the approach taken here is fairly simplistic in that records where VISIT='Baseline' are flagged as the baseline record. In practice, this might not be the case. For example, it might be that a given patient does not have a VISIT= 'Baseline' record, and you need to go to that patient's screening visit to pull an appropriate baseline record for flagging. For reasons like this, some in the industry would argue that the flagging of baseline records is a task best left to the analysis datasets in ADaM.

❺ At this point, the LB file is mostly built, and now we want to create the LBDY variable. To create LBDY, we must compare the dates stored in variables RFSTDTC and LBDTC. The DM and draft LB datasets are merged, and the %make_sdtm_dy macro is called to define LBDY.

❻ This DATA step is designed to create LBSEQ. There are some interesting features in this step. First, we need to DROP the old LBSEQ variable and re-create it because of the inherent data retention feature of the SAS DATA step. Because we have to drop and re-create the LBSEQ variable, we need that RETAIN statement to ensure that the variables appear in LB from left to right in the expected order. We also need the LABEL statement to redefine the label for LBSEQ. The warning message in that DATA step is there to ensure the uniqueness of the SDTM record. Keep in mind that although --SEQ and KEYSEQUENCE (found in the VARIABLE_METADATA metadata tab) are probably analogous for any domain, they do not have to be. It is very possible that the KEYSEQUENCE list of variables is not granular enough or specific enough to define --SEQ. Because each value of --SEQ must identify a unique record, it might require a more comprehensive list of variables than what is found in KEYSEQUENCE.

❼ The LB program ends with getting the prescribed sort order from the %make_sort_order macro and saving the final LB dataset. The final dataset is sorted and contains only the SDTM variables that we want.

Building a Custom XP Findings Domain

The following SAS code builds the XP dataset from our source clinical trial data found in "Appendix A - Source Data Programs." Because there is no "headache pain" domain already defined in the SDTM, XP is a user-generated domain designed to capture the pain efficacy measurements for this study. For more information about how to build customized user-written domains, see Section 2.6 of the "CDISC SDTM Implementation Guide." It is worth noting that there are new therapeutic area CDISC standards that extend the general SDTM into specific disease implementations, and there is in fact a draft pain assessment SDTM standard. However, that standard doesn't specifically address pain intensity, and we wanted to illustrate the creation of a custom domain here, so the pain standard is not used here. The following program assumes that the source datasets can be found under the libref `source` and that the permanent SAS formats can be found under the libref `library`.

```
*------------------------------------------------------------*;
* XP.sas creates the SDTM XP dataset and saves it
* as a permanent SAS datasets to the target libref.
*------------------------------------------------------------*;

**** CREATE EMPTY DM DATASET CALLED EMPTY_XP;
%make_empty_dataset(metadatafile=SDTM_METADATA.xls,dataset=XP)   ❶

proc format;
  value pain
    0='None'
    1='Mild'
    2='Moderate'
    3='Severe';
run;

**** DERIVE THE MAJORITY OF SDTM XP VARIABLES;
options missing = ' ';
data xp;
  set EMPTY_XP
      source.pain;

    studyid = 'XYZ123';
    domain = 'XP';
    usubjid = left(uniqueid);

    xptest = 'Pain Score';                      ❷
    xptestcd = 'XPPAIN';
    epoch = 'TREATMENT';

    **** transpose pain data;                    ❸
    array dates {3} randomizedt month3dt month6dt;
    array scores {3} painbase pain3mo pain6mo;

    do i = 1 to 3;
      visitnum = i - 1;
      visit = put(visitnum,visit_labs_month.);
      if visit = 'Baseline' then
        xpblfl = 'Y';
      else
       xpblfl = ' ';

      if scores{i} ne . then
        do;
          xporres = left(put(scores{i},pain.));
          xpstresc = xporres;
          xpstresn = scores{i};
```

```
            xpdtc = put(dates{i},yymmdd10.);
            output;
         end;
      end;
run;


proc sort
  data=xp;
    by usubjid;
run;

**** CREATE SDTM STUDYDAY VARIABLES;                    ❹
data xp;
  merge xp(in=inxp) target.dm(keep=usubjid rfstdtc);
    by usubjid;

    if inxp;

    %make_sdtm_dy(date=xpdtc)
run;


**** CREATE SEQ VARIABLE;                               ❺
proc sort
  data=xp;
    by studyid usubjid xptestcd visitnum;
run;


data xp;
  retain &XPKEEPSTRING;
  set xp(drop=xpseq);
    by studyid usubjid xptestcd visitnum;

  if not (first.visitnum and last.visitnum) then
   put "WARN" "ING: key variables do not define an unique record. "
         usubjid=;

    retain xpseq;
    if first.usubjid then
      xpseq = 1;
    else
      xpseq = xpseq + 1;

    label xpseq = "Sequence Number";
run;

**** SORT XP ACCORDING TO METADATA AND SAVE PERMANENT DATASET;   ❻
%make_sort_order(metadatafile=SDTM_METADATA.xls,dataset=XP)

proc sort
  data=xp(keep = &XPKEEPSTRING)
  out=target.xp;
    by &XPSORTSTRING;
run;
```

When the XP.sas program has been run, the XP SDTM domain dataset is saved to the target libref.

❶ The first step in this program creates the empty SDTM XP dataset called EMPTY_XP based on the domain metadata spreadsheet. EMPTY_XP is then set with the source `pain` dataset, and variable population begins.

❷ Because there is only one measurement of Pain Score in this data, XPTEST and XPTESTCD can have simple assignment statements here.

❸ This section of the DATA step transposes the data so that there is one record output per visit.

❹ At this point, the XP file is mostly built. Now we want to create the XPDY variable. To create XPDY, we must compare the variables RFSTDTC and XPDTC. The DM and draft XP datasets are merged, and the %make_sdtm_dy macro is called to define XPDY.

❺ XPSEQ is created at this point.

❻ The XP program ends with getting the prescribed sort order from the %make_sort_order macro and saving the final XP dataset. The final dataset is sorted and contains only the SDTM variables that we want.

Building the AE Events Domain

The following SAS code builds the AE adverse events domain dataset from our source clinical trial data found in "Appendix A - Source Data Programs." The following program assumes that the source datasets can be found under the libref `source` and that the permanent SAS formats can be found under the libref `library`.

```
*--------------------------------------------------------------*;
* AE.sas creates the SDTM AE dataset and saves it
* as permanent SAS datasets to the target libref.
*--------------------------------------------------------------*;

**** CREATE EMPTY DM DATASET CALLED EMPTY_AE;                    ❶
%make_empty_dataset(metadatafile=SDTM_METADATA.xls,dataset=AE)


**** DERIVE THE MAJORITY OF SDTM AE VARIABLES;
options missing = ' ';
data ae;
  set EMPTY_AE
  source.adverse(rename=(aerel=_aerel aesev=_aesev));

    studyid = 'XYZ123';
    domain = 'AE';
    usubjid = left(uniqueid);
    aeterm = left(aetext);
    aedecod = left(prefterm);
    aeptcd = left(ptcode);
    aellt = left(llterm);
    aelltcd = left(lltcode);
    aehlt = left(hlterm);
    aehltcd = left(hltcode);
    aehlgt = left(hlgterm);
    aehlgtcd = left(hlgtcod);
    aesoc = left(bodysys);
    aebodsys = aesoc;
```

```
      aebdsycd = left(soccode);
      aesoccd = aebdsycd;

      aesev = put(_aesev,aesev_adverse_aesev.);            ❷
      aeacn = put(aeaction,acn_adverse_aeaction.);
      aerel = put(_aerel,aerel_adverse_aerel.);
      aeser = put(serious,$ny_adverse_serious.);
      aestdtc = put(aestart,yymmdd10.);
      aeendtc = put(aeend,yymmdd10.);
      epoch = 'TREATMENT';
      if aeser = 'Y' then
         aeslife = 'Y';
run;


proc sort
   data=ae;
      by usubjid;
run;

**** CREATE SDTM STUDYDAY VARIABLES;                       ❸
data ae;
   merge ae(in=inae) target.dm(keep=usubjid rfstdtc);
      by usubjid;

      if inae;

      %make_sdtm_dy(date=aestdtc);
      %make_sdtm_dy(date=aeendtc);
run;


**** CREATE SEQ VARIABLE;                                  ❹
proc sort
   data=ae;
      by studyid usubjid aedecod aestdtc aeendtc;
run;

data ae;
   retain &AEKEEPSTRING;
   set ae(drop=aeseq);
      by studyid usubjid aedecod aestdtc aeendtc;

   if not (first.aeendtc and last.aeendtc) then
    put "WARN" "ING: key variables do not define an unique record. "
          usubjid=;

    retain aeseq 0;
    aeseq = aeseq + 1;

    label aeseq = "Sequence Number";
run;


**** SORT AE ACCORDING TO METADATA AND SAVE PERMANENT DATASET; ❺
%make_sort_order(metadatafile=SDTM_METADATA.xls,dataset=AE)
```

```
proc sort
  data=ae(keep = &AEKEEPSTRING)
  out=target.ae;
    by &AESORTSTRING;
run;
```

When the AE.sas program has been run, the AE SDTM domain dataset is saved to the target libref.

❶ The first step in this program creates the empty SDTM AE dataset called EMPTY_AE based on the domain metadata spreadsheet.

❷ Note here that these are SAS formats that are created by the make_codelist_formats.sas program and that hold the controlled terminology for AESEV, AEACN, AEREL, and AESER. Because the AESEV and AEREL variables existed already in the source adverse dataset, they had to be renamed in the SET statement.

❸ The %make_sdtm_dy macro is called twice here to create AESTDY and AEENDY.

❹ The sequence variable AESEQ is created in a fashion similar to what was done for previous domains.

❺ The program calls %make_sort_order here and saves the permanent domain dataset as was done before.

Building the EX Exposure Interventions Domain

The following SAS code builds the EX exposure domain dataset from our source clinical trial data found in "Appendix A - Source Data Programs." The following program assumes that the source datasets can be found under the libref `source` and that the permanent SAS formats can be found under the libref `library`.

```
*----------------------------------------------------------------*;
* EX.sas creates the SDTM EX dataset and saves it
* as a permanent SAS datasets to the target libref.
*----------------------------------------------------------------*;

**** CREATE EMPTY EX DATASET CALLED EMPTY_EX;
%make_empty_dataset(metadatafile=VARIABLE_METADATA.xlsx,dataset=EX)

**** DERIVE THE MAJORITY OF SDTM EX VARIABLES;
options missing = ' ';
data ex;
  set EMPTY_EX
      source.dosing;

    studyid = 'XYZ123';
    domain = 'EX';
    usubjid = left(uniqueid);
    exdose = dailydose;
    exdostot = dailydose;
    exdosu = 'mg';
    exdosfrm = 'TABLET, COATED';
    %make_dtc_date(dtcdate=exstdtc, year=startyy, month=startmm, ❶
                   day=startdd)
```

```
      %make_dtc_date(dtcdate=exendtc, year=endyy, month=endmm,
                   day=enddd)
run;


proc sort
  data=ex;
    by usubjid;
run;


**** CREATE SDTM STUDYDAY VARIABLES AND INSERT EXTRT;
data ex;
  merge ex(in=inex) target.dm(keep=usubjid rfstdtc arm);
    by usubjid;

    if inex;

    %make_sdtm_dy(date=exstdtc);
    %make_sdtm_dy(date=exendtc);

    **** in this simplistic case all subjects received the
    **** treatment they were randomized to;              ❷
    extrt = arm;
run;


**** CREATE SEQ VARIABLE;
proc sort
  data=ex;
    by studyid usubjid extrt exstdtc;
run;

data ex;
  retain &EXKEEPSTRING;
  set ex(drop=exseq);
    by studyid usubjid extrt exstdtc;

    if not (first.exstdtc and last.exstdtc) then
      put "WARN" "ING: key variables do not define an unique"
         " record. " usubjid=;

    retain exseq;
    if first.usubjid then
      exseq = 1;
    else
      exseq = exseq + 1;

    label exseq = "Sequence Number";
run;


**** SORT EX ACCORDING TO METADATA AND SAVE PERMANENT DATASET;
%make_sort_order(metadatafile=TOC_METADATA.xlsx,dataset=EX)
```

```
proc sort
  data=ex(keep = &EXKEEPSTRING)
  out=target.ex;
    by &EXSORTSTRING;
run;
```

When the EX.sas program has been run, the EX SDTM domain dataset is saved to the target libref. Similar methods that were seen in previous domains were used here to create the EX domain.

❶ To keep things a bit simple and clean, our date sources have been complete dates so far. At this point, we use the %make_dtc_date SAS macro that creates EXSTDTC and EXENDTC from partial dates. If you look at the patient with USUBJID="UNI712" as the first record, the patient has only a start year, so EXSTDTC="2010" and EXSTDY are missing. For the second record, the patient has both a year and month present for the dosing stop date, so EXENDTC="2010-12" and EXENDY are missing.

❷ The way that EXTRT is derived here is perhaps overly simplistic. EXTRT should contain the name of the actual treatment that the subject received. Here we assume that the subject received what they were randomized to, which is present in the DM ARM variable.

Building Trial Design Model (TDM) Domains

The following SAS code builds the TA, TD, TE, TI, TS, and TV datasets. These TDM domains consist entirely of trial metadata and contain no actual patient data. Because this is the case, you will not typically find this data in your underlying clinical data management systems. For this chapter, this metadata was entered manually into an Excel file called trialdesign.xlsx. Until clinical trial data systems mature significantly, you might find that you have to create this study metadata manually as well. See Chapter 13 for a discussion of the related Protocol Representation Model. There are two other TDM datasets that are based on the subject clinical data called SE and SV, but those are not presented or created here.

```
*-------------------------------------------------------------*;
* TDM.sas creates the SDTM TA, TD, TE, TI, TS, and TV datasets and
* saves them as a permanent SAS datasets to the target libref.
*-------------------------------------------------------------*;

**** CREATE EMPTY TA DATASET CALLED EMPTY_TA;                    ❶
%make_empty_dataset(metadatafile=SDTM_METADATA.xls,dataset=TA)

proc import                                                      ❷
  datafile="trialdesign.xls"
  out=ta
  dbms=excelcs
  replace;
  sheet='TA';
run;

**** SET EMPTY DOMAIN WITH ACTUAL DATA;                          ❸
data ta;
  set EMPTY_TA
      ta;
run;
```

```
**** SORT DOMAIN ACCORDING TO METADATA AND SAVE PERMANENT DATASET;
%make_sort_order(metadatafile=SDTM_METADATA.xls,dataset=TA)        ❹

proc sort
  data=ta(keep = &TAKEEPSTRING)
  out=target.ta;
    by &TASORTSTRING;
run;


**** CREATE EMPTY TD DATASET CALLED EMPTY_TD;
%make_empty_dataset(metadatafile=SDTM_METADATA.xls,dataset=TD)

proc import
  datafile="trialdesign.xls"
  out=td
  dbms=excelcs
  replace;
  sheet='TD';
run;

**** SET EMPTY DOMAIN WITH ACTUAL DATA;
data td;
  set EMPTY_TD
      td;
run;

**** SORT DOMAIN ACCORDING TO METADATA AND SAVE PERMANENT DATASET;
%make_sort_order(metadatafile=SDTM_METADATA.xls,dataset=TD)

proc sort
  data=td(keep = &TDKEEPSTRING)
  out=target.td;
    by &TDSORTSTRING;
run;


**** CREATE EMPTY TE DATASET CALLED EMPTY_TE;
%make_empty_dataset(metadatafile=SDTM_METADATA.xls,dataset=TE)

proc import
  datafile="trialdesign.xls"
  out=te
  dbms=excelcs
  replace;
  sheet='TE';
run;

**** SET EMPTY DOMAIN WITH ACTUAL DATA;
data te;
  set EMPTY_TE
      te;
run;

**** SORT DOMAIN ACCORDING TO METADATA AND SAVE PERMANENT DATASET;
%make_sort_order(metadatafile=SDTM_METADATA.xls,dataset=TE)
```

```
proc sort
  data=te(keep = &TEKEEPSTRING)
  out=target.te;
    by &TESORTSTRING;
run;



**** CREATE EMPTY TI DATASET CALLED EMPTY_TI;
%make_empty_dataset(metadatafile=SDTM_METADATA.xls,dataset=TI)

proc import
  datafile="trialdesign.xls"
  out=ti
  dbms=excelcs
  replace;
  sheet='TI';
run;

**** SET EMPTY DOMAIN WITH ACTUAL DATA;
data ti;
  set EMPTY_TI
      ti;
run;

**** SORT DOMAIN ACCORDING TO METADATA AND SAVE PERMANENT DATASET;
%make_sort_order(metadatafile=SDTM_METADATA.xls,dataset=TI)

proc sort
  data=ti(keep = &TIKEEPSTRING)
  out=target.ti;
    by &TISORTSTRING;
run;



**** CREATE EMPTY TS DATASET CALLED EMPTY_TS;
%make_empty_dataset(metadatafile=SDTM_METADATA.xls,dataset=TS)

proc import
  datafile="trialdesign.xls"
  out=ts
  dbms=excelcs
  replace;
  sheet='TS';
run;

**** SET EMPTY DOMAIN WITH ACTUAL DATA;
data ts;
  set EMPTY_TS
      ts;
run;

**** SORT DOMAIN ACCORDING TO METADATA AND SAVE PERMANENT DATASET;
%make_sort_order(metadatafile=SDTM_METADATA.xls,dataset=TS)
```

```
proc sort
  data=ts(keep = &TSKEEPSTRING)
  out=target.ts;
    by &TSSORTSTRING;
run;



**** CREATE EMPTY TV DATASET CALLED EMPTY_TV;
%make_empty_dataset(metadatafile=SDTM_METADATA.xls,dataset=TV)

proc import
  datafile="trialdesign.xls"
  out=tv
  dbms=excelcs
  replace;
  sheet='TV';
run;

**** SET EMPTY DOMAIN WITH ACTUAL DATA;
data tv;
  set EMPTY_TV
      tv;
run;

**** SORT DOMAIN ACCORDING TO METADATA AND SAVE PERMANENT DATASET;
%make_sort_order(metadatafile=SDTM_METADATA.xls,dataset=TV)

proc sort
  data=tv(keep = &TVKEEPSTRING)
  out=target.tv;
    by &TVSORTSTRING;
run;
```

When the TDM.sas program has been run, the TA, TE, TI, TS, and TV SDTM domain datasets are saved to the target libref.

❶ For each of the TDM domains, the %make_empty_dataset macro is run to create the empty dataset for populating.

❷ PROC IMPORT is used to import the trial design dataset metadata spreadsheet from Excel.

❸ The empty domain dataset structure is set with the actual data here.

❹ The proper sort order is obtained from the %make_sort_order macro, and then the permanent domain dataset is stored.

Chapter Summary

The best approach to implementing the SDTM in Base SAS is through the heavy use of metadata and SAS macros for repetitive task automation.

Build a set of SAS macros and programs that handle SDTM domain construction repetitive tasks. Some examples in this chapter include an empty dataset creator, a dataset sorter, a --DTC date creator, a study day calculator, and an automated format library generator. You can add more SAS macro-based tools as you see fit.

Keep in mind that the examples in this chapter were made somewhat simple to keep the size of the book manageable and easily understood. When you convert data to the SDTM, it can get fairly complicated fairly fast, so the examples in this chapter might provide only superficial treatment to some complicated SDTM data conversion issue. This chapter assumes that standardized lab results are the same as original results, that there is one record per nominal visit, that defining baseline flags --BLFL is simple, and generally that the incoming data is somewhat well behaved, which is rarely the case in clinical study data.

Chapter 4: Implementing CDISC SDTM with the SAS Clinical Standards Toolkit and Base SAS

In Chapter 3, you saw how to create CDISC SDTM data just using Base SAS and metadata stored in Microsoft Excel files. In this chapter, you will explore another approach where you replace Microsoft Excel with the SAS Clinical Standards Toolkit as a metadata repository. You will first migrate the SDTM metadata from Excel into SAS Clinical Standards Toolkit 1.6. The metadata in the SAS Clinical Standards Toolkit and some SAS macros provided as part of that tool will be used to build the SDTM datasets. Finally, the SAS Clinical Standards Toolkit will be used to build define.xml.

More Information

Appendix A - Source Data Programs

SAS Clinical Standards Toolkit 1.6: User's Guide

SAS Clinical Standards Toolkit Background

SAS created the SAS Clinical Standards Toolkit in response to the pharmaceutical industry's need for a way to store and leverage clinical study metadata. The SAS Clinical Standards Toolkit is, at its core, a set of SAS datasets representing metadata, some text configuration files, and a set of SAS macros that work with those datasets and configuration files. A very nice plus to the SAS Clinical Standards Toolkit is the price. It is available to those with a Base SAS license (SAS 9.2 and later, and SAS 9.1.3 via a hot fix) by making a special request of your SAS sales representative. Another nice plus is that if you already understand Base SAS and the SAS macro language, then you have the basic technical understanding for how the SAS Clinical Standards Toolkit works.

The examples in this chapter are built in SAS Clinical Standards Toolkit 1.6. We expect that this toolset will evolve over time and as the CDISC models also evolve. You can find the documentation for the toolkit and for what was done in this chapter in the *SAS Clinical Standards Toolkit 1.6: User's Guide.*

Clinical Standards Setup for Study XYZ123

In this section, you will see how to define the SDTM data for clinical study XYZ123 in the SAS Clinical Standards Toolkit. It is assumed that you have the SAS Clinical Standards Toolkit installed, and that you know where your CST global standards library resides. For the examples below, the CST global standards library is found at C:\cstGlobalLibrary, but your installation directory might differ. If you have installed the CST to a different location, you need to replace this specified path below with your installation path.

Copy SDTM 3.2 Standard to XYZ123

A common use case for many people is to define an SDTM standard based on the CDISC SDTM. That is what we want to do here. So copy the SDTM 3.2 CST-provided metadata over to a new study-specific XYZ123 area. For this example, copy the entire contents of C:\cstGlobalLibrary\standards\cdisc-sdtm-3.2-1.6 over to C:\cstGlobalLibrary\standards\XYZ_program_cdisc-sdtm-3.2-1.6.

Edit Study XYZ123 Metadata

To customize the SDTM XYZ123 metadata and make it differ from the generic CST-supplied CDISC SDTM 3.2 metadata, you need to edit some metadata datasets. For this example, all edits to SAS datasets were done by editing the datasets directly within SAS.

First, you need to update the domain-level metadata, which can be found in C:\cstGlobalLibrary\standards\XYZ_program_cdisc-sdtm-3.2-1.6\metadata\reference_tables.sas7bdat. Because SUPPDM and the XYZ123 custom domain called XP are added to the 3.2 standard, you have to add two rows to the reference_tables dataset to look like this:

	SASref	Table	Label	Class	XmlPath	XmlTitle	Structure	Purpose
48	REFMETA	SUPPDM	DM - Supplemental Qualifiers	Special Purpose Datasets	/transport/suppdm.xpt	DM - Supplemental Qualifiers SAS transport file	One record per IDVAR, IDVARVAL, and QNAM value per subject	Tabulation
59	REFMETA	XP	Pain Scores	Findings	/transport/xp.xpt	Pain Scores SAS transport file	One record per subject per visit	Tabulation

	Keys	State	Date	Standard	StandardVersion	Standardref	comment
48	STUDYID RDOMAIN USUBJID IDVAR IDVARVAL QNAM	Final	2013-11-26	CDISC-SDTM	XYZ-SDTM-3.2		
59	STUDYID USUBJID XPTESTCD VISITNUM	Final	2013-11-26	CDISC-SDTM	XYZ-SDTM-3.2		

The standardversion column for all other rows/domains is also changed to "XYZ-SDTM-3.2" to point at the newly revised standard. Also note that the comment column here is blank. The CST is shipped with this column populated with the CDISC Notes column from the SDTM standard. But those note entries can cause the CST to crash due to special formatting characters loaded from the Microsoft Excel CDISC metadata file. So it is advisable that you delete that text or replace it with text that makes sense for your organization.

Now that the study and domain-level information has been defined, you need to edit the domain variable-level metadata. That dataset can be found at C:\cstGlobalLibrary\standards\XYZ_program_cdisc-sdtm-3.2-

1.6\metadata\reference_columns.sas7bdat. Here again, you need to add the data for the new XP and SUPPDM domains. The standardversion, order, length, displayformat, xmlcodelist, origin, term, and algorithm columns were edited as needed. Several of the domains were edited to remove SDTM variables that were optional. The other domains not used specifically in the conversion were left as they were. Lengths on studyid, domain, and usubjid were updated for consistency across domains, even though a majority of the domains are not used in this exercise.

Note that editing the reference_columns dataset can be quite time-consuming. This is the step where you have to enter the majority of your standard metadata. CDISC does not specify standard lengths of variables for the SDTM, so you need to make that effort here in the length variable. Order needs to be set so that you have the variables appear in your dataset in the proper order. The codelist needs to be set to point to the proper controlled terminology. Origin is set to tell the reviewer where the variable can be found on the CRF. The algorithm column defines your computational method if you have one. Here is a snapshot of the edits made to the DM rows of the reference_columns dataset:

	SASref	table	column	label	order	type	length	displayformat	xmldatatype	xmlcodelist	core	origin	role	term	algorithm	qualifiers	standard	standardversion
193	REFMETA	DM	STUDYID	Study Identifier	1	C	15		text		Req		Identifier			UPPERCASE	CDISC-SDTM	XYZ-SDTM-3.2
194	REFMETA	DM	DOMAIN	Domain Abbreviation	2	C	2		text		Req		Identifier	DM		UPPERCASE	CDISC-SDTM	XYZ-SDTM-3.2
195	REFMETA	DM	USUBJID	Unique Subject Identifier	3	C	25		text		Req		Identifier			UPPERCASE	CDISC-SDTM	XYZ-SDTM-3.2
196	REFMETA	DM	SUBJID	Subject Identifier for the Study	4	C	7		text		Req		Topic			UPPERCASE	CDISC-SDTM	XYZ-SDTM-3.2
197	REFMETA	DM	RFSTDTC	Subject Reference Start Date/Time	5	C	16		datetime		Exp		RecordQualifier	ISO 8601		UPPERCASE DATETIME	CDISC-SDTM	XYZ-SDTM-3.2
198	REFMETA	DM	RFENDTC	Subject Reference End Date/Time	6	C	16		datetime		Exp		RecordQualifier	ISO 8601		UPPERCASE DATETIME	CDISC-SDTM	XYZ-SDTM-3.2
199	REFMETA	DM	RFXSTDTC	Date/Time of First Study Treatment	7	C	16		datetime		Exp		RecordQualifier	ISO 8601		UPPERCASE DATETIME	CDISC-SDTM	XYZ-SDTM-3.2
200	REFMETA	DM	RFXENDTC	Date/Time of Last Study Treatment	8	C	16		datetime		Exp		RecordQualifier	ISO 8601		UPPERCASE DATETIME	CDISC-SDTM	XYZ-SDTM-3.2
201	REFMETA	DM	RFICDTC	Date/Time of Informed Consent	9	C	16		datetime		Exp		RecordQualifier	ISO 8601		UPPERCASE DATETIME	CDISC-SDTM	XYZ-SDTM-3.2
202	REFMETA	DM	RFPENDTC	Date/Time of End of Participation	10	C	16		datetime		Exp		RecordQualifier	ISO 8601		UPPERCASE DATETIME	CDISC-SDTM	XYZ-SDTM-3.2
203	REFMETA	DM	DTHDTC	Date/Time of Death	11	C	16		datetime		Exp		RecordQualifier	ISO 8601		UPPERCASE DATETIME	CDISC-SDTM	XYZ-SDTM-3.2
204	REFMETA	DM	DTHFL	Subject Death Flag	12	C	2		text	NY	Exp		RecordQualifier	(NY)		UPPERCASE	CDISC-SDTM	XYZ-SDTM-3.2
205	REFMETA	DM	SITEID	Study Site Identifier	13	C	7		text		Req		RecordQualifier			UPPERCASE	CDISC-SDTM	XYZ-SDTM-3.2
206	REFMETA	DM	BRTHDTC	Date/Time of Birth	14	C	16		datetime		Perm		RecordQualifier	ISO 8601		UPPERCASE DATETIME	CDISC-SDTM	XYZ-SDTM-3.2
207	REFMETA	DM	AGE	Age	15	N	8	8.1	float		Exp		RecordQualifier				CDISC-SDTM	XYZ-SDTM-3.2
208	REFMETA	DM	AGEU	Age Units	16	C	10		text	AGEU	Exp		VariableQualifier	(AGEU)		UPPERCASE	CDISC-SDTM	XYZ-SDTM-3.2
209	REFMETA	DM	SEX	Sex	17	C	2		text	SEX	Req		RecordQualifier	(SEX)		UPPERCASE	CDISC-SDTM	XYZ-SDTM-3.2
210	REFMETA	DM	RACE	Race	18	C	80		text	RACE	Exp		RecordQualifier	(RACE)		UPPERCASE	CDISC-SDTM	XYZ-SDTM-3.2
211	REFMETA	DM	ARMCD	Planned Arm Code	19	C	8		text		Req		RecordQualifier	*		UPPERCASE	CDISC-SDTM	XYZ-SDTM-3.2
212	REFMETA	DM	ARM	Description of Planned Arm	20	C	40		text		Req		SynonymQualifier	*		UPPERCASE	CDISC-SDTM	XYZ-SDTM-3.2
213	REFMETA	DM	ACTARMCD	Actual Arm Code	21	C	8		text		Req		RecordQualifier	*		UPPERCASE	CDISC-SDTM	XYZ-SDTM-3.2
214	REFMETA	DM	ACTARM	Description of Actual Arm	22	C	40		text		Req		SynonymQualifier	*		UPPERCASE	CDISC-SDTM	XYZ-SDTM-3.2
215	REFMETA	DM	COUNTRY	Country	23	C	3		text	COUNTRY	Req		RecordQualifier	(COUNTRY) ISO 3166		UPPERCASE	CDISC-SDTM	XYZ-SDTM-3.2

Edit the XYZ123 SAS Clinical Standards Toolkit Control Files

Now that the XYZ123 metadata has been customized in the SAS Clinical Standards Toolkit to meet your needs, you need to make some edits to a few control files so that you can register the XYZ123 standard into the toolkit. First, you need to copy the initialization files initialize.properties and validation.properties from the base SDTM standard at C:\cstGlobalLibrary\standards\cdisc-sdtm-3.2-1.6\programs to the XYZ custom standard at C:\cstGlobalLibrary\standards\XYZ_program_cdisc-sdtm-3.2-1.6\programs.

The first control metadata dataset to be edited is the standards dataset that holds the study-level metadata found at C:\cstGlobalLibrary\standards\XYZ_program_cdisc-sdtm-3.2-1.6\control\standards.sas7bdat. The standardversion, mnemonic, comment, and rootpath variables were edited to look like this:

	standard	mnemonic	standardversion	groupname	groupversion	comment	rootpath	studylibraryrootpath
1	CDISC-SDTM	XYZS	XYZ-SDTM-3.2	SDTM	3.2	XYZprogram standard based on CDISC SDTM V3.2	&_cstGRoot/standards/XYZ_program_cdisc-sdtm-3.2-1.6	&_cstGRoot/standards/XYZ_program_cdisc-sdtm-3.2-1.6

	controlsubfolder	templatesubfolder	isstandarddefault	iscstframework	isdatastandard	supportsvalidation	isxmlstandard	importxsl	exportxsl	schema	productrevision
1	control	templates	Y	N	Y	Y	N				1.6

Next, you need to edit the standardsasreferences control dataset for the registration program to run. The standardsasreferences dataset is a controlling metadata dataset that tells the SAS Clinical Standards Toolkit where it can find the objects that it needs for the various tasks that it performs. The standardsasreferences dataset for standard XYZ123 can be found at C:\cstGlobalLibrary\standards\XYZ_program_cdisc-sdtm-3.2-

1.6\control\standardsasreferences.sas7bdat. Change the standard variable to XYZ std CDISC-SDTM. The resulting standardsasreferences dataset for this job looks like this:

	standard	standardversion	type	subtype	SASref	reftype	iotype	filetype	allowoverwrite	relpathprefix	path	order	memname	comment
1	CDISC-SDTM	XYZ-SDTM-3.2	autocall		autocall	fileref	input	folder	N	rootpath	macros	1		
2	CDISC-SDTM	XYZ-SDTM-3.2	classmetadata	column	refmeta	libref	input	dataset	N	rootpath	metadata		class_columns.sas7bdat	
3	CDISC-SDTM	XYZ-SDTM-3.2	classmetadata	table	refmeta	libref	input	dataset	N	rootpath	metadata		class_tables.sas7bdat	
4	CDISC-SDTM	XYZ-SDTM-3.2	cstmetadata	lookup	stdmeta	libref	input	dataset	N	rootpath	control		standardlookup.sas7bdat	Standard-specific lookup data set
5	CDISC-SDTM	XYZ-SDTM-3.2	cstmetadata	macrovariabledetails	stdmeta	libref	input	dataset	N	rootpath	control		standardmacrovariabledetails.sas7bdat	Standard-specific macro variable details data set
6	CDISC-SDTM	XYZ-SDTM-3.2	cstmetadata	macrovariables	stdmeta	libref	input	dataset	N	rootpath	control		standardmacrovariables.sas7bdat	Standard-specific macro variables data set
7	CDISC-SDTM	XYZ-SDTM-3.2	cstmetadata	sasreferences	stdmeta	libref	input	dataset	N	rootpath	control		standardsasreferences.sas7bdat	Standard-specific SASReferences data set
8	CDISC-SDTM	XYZ-SDTM-3.2	cstmetadata	standard	stdmeta	libref	input	dataset	N	rootpath	control		standards.sas7bdat	Standard-specific standards data set
9	CDISC-SDTM	XYZ-SDTM-3.2	lookup		lookup	libref	input	dataset	N	rootpath	control		standardlookup.sas7bdat	
10	CDISC-SDTM	XYZ-SDTM-3.2	messages		messages	libref	input	dataset	N	rootpath	messages	1	messages.sas7bdat	
11	CDISC-SDTM	XYZ-SDTM-3.2	properties	initialize	initprop	fileref	input	file	N	rootpath	programs	1	initialize.properties	Initialization properties when using the standard
12	CDISC-SDTM	XYZ-SDTM-3.2	properties	validation	valprop	fileref	input	file	N	rootpath	programs	2	validation.properties	Sets up default properties used in validation
13	CDISC-SDTM	XYZ-SDTM-3.2	referencecontrol	checktable	refcntl	libref	input	dataset	N	rootpath	validation/control		validation_domainsbycheck.sas7bdat	
14	CDISC-SDTM	XYZ-SDTM-3.2	referencecontrol	standardref	refcntl	libref	input	dataset	N	rootpath	validation/control		validation_stdref.sas7bdat	
15	CDISC-SDTM	XYZ-SDTM-3.2	referencecontrol	validation	refcntl	libref	input	dataset	N	rootpath	validation/control		validation_master.sas7bdat	
16	CDISC-SDTM	XYZ-SDTM-3.2	referencemetadata	column	refmeta	libref	input	dataset	N	rootpath	metadata		reference_columns.sas7bdat	
17	CDISC-SDTM	XYZ-SDTM-3.2	referencemetadata	table	refmeta	libref	input	dataset	N	rootpath	metadata		reference_tables.sas7bdat	
18	CDISC-SDTM	XYZ-SDTM-3.2	template		tmplt	libref	input	folder	N	rootpath	templates			

You also need to edit the standardelookup control dataset for the registration program to run. The standardlookup dataset for standard XYZ123 can be found at C:\cstGlobalLibrary\standards\XYZ_program_cdisc-sdtm-3.2-1.6\control\standardlookup.sas7bdat. Primarily the standard and standardversion columns need to be modified to point to the XYZ123 standard. A few rows of this dataset look like this:

	standard	standardversion	SASref	table	column	refcolumn	refvalue	value	default	nonnull	order	templatetype	template	comment
1	CDISC-SDTM	XYZ-SDTM-3.2	refcntl	validation_master	checkseverity			NOTE	Y	N	1			
2	CDISC-SDTM	XYZ-SDTM-3.2	refcntl	validation_master	checkseverity			WARNING	N	N	2			
3	CDISC-SDTM	XYZ-SDTM-3.2	refcntl	validation_master	checkseverity			ERROR	N	N	3			
4	CDISC-SDTM	XYZ-SDTM-3.2	refcntl	validation_master	checksource			SAS	Y	Y	1			
5	CDISC-SDTM	XYZ-SDTM-3.2	refcntl	validation_master	checksource			JANUS	N	Y	2			
6	CDISC-SDTM	XYZ-SDTM-3.2	refcntl	validation_master	checksource			JANUSFR	N	Y	3			
7	CDISC-SDTM	XYZ-SDTM-3.2	refcntl	validation_master	checksource			WEBSDM	N	Y	4			
8	CDISC-SDTM	XYZ-SDTM-3.2	refcntl	validation_master	checksource			OPENCDISC	N	Y	5			
9	CDISC-SDTM	XYZ-SDTM-3.2	refcntl	validation_master	checktype			METADATA	Y	N	1			
10	CDISC-SDTM	XYZ-SDTM-3.2	refcntl	validation_master	checktype			COLUMNATTRIBUTE	N	N	2			
11	CDISC-SDTM	XYZ-SDTM-3.2	refcntl	validation_master	checktype			COLUMNVALUE	N	N	3			
12	CDISC-SDTM	XYZ-SDTM-3.2	refcntl	validation_master	checktype			DATE	N	N	4			

It is critical to get these control datasets right in order to have your SAS Clinical Standards Toolkit jobs run properly. The *SAS Clinical Standards Toolkit: User's Guide* has much more detail on the content of these datasets, and you should refer there for extensive details.

Register the XYZ123 Standard

You now need to run a program to register the XYZ123 SDTM standard into the SAS Clinical Standards Toolkit. Copy the registerstandard.sas program from the base SDTM 3.2 area C:\cstGlobalLibrary\standards\cdisc-sdtm-3.2-1.6\programs over to the XYZ123 standard area at C:\cstGlobalLibrary\standards\XYZ_program_cdisc-sdtm-3.2-1.6\programs. You can use that program as a basis for your registration program. In this example, the start of the final registration program registerstandard.sas looks like this:

```
* Register the XYZ program standard to the global library;

* Register the standard to the global library;

* Set up macro variables needed by this program;          ❶
%let _thisStandard=CDISC-SDTM;
%let _thisStandardVersion=XYZ-SDTM-3.2;
%let _thisProductRevision=1.6;
%let _thisDirWithinStandards=XYZ_program_cdisc-sdtm-3.2-1.6;

%cstutil_setcstgroot;                                      ❷
```

```
* Set the framework properties used for the uninstall;      ❸

%cst_setStandardProperties(
  _cstStandard=CST-FRAMEWORK,
  _cstSubType=initialize
  );
❹
```

When the registerstandard.sas program is run, the XYZ123 standard becomes available for use by the SAS Clinical Standards Toolkit.

❶ You need to define these four macro variables. Beyond this, the code is standard CST program code.

❷ This step sets the &_cstgroot macro variable used by the code below it.

❸ This step initializes the toolkit environment.

❹ From this point, the program calls the %cst_registerStandard toolkit SAS macro that puts all of the XYZ123 standard metadata into the toolkit.

Now that the XYZ123 standard has been defined, this custom SDTM standard is available for use in Base SAS in order to convert the source data to SDTM datasets. In Chapter 3, you used Excel to store your study metadata, but here you will use the metadata in SAS Clinical Standards Toolkit. Also, the work that you have done here will be leveraged in Chapter 5 when you use this custom XYZ123 SDTM standard defined in the SAS Clinical Standards Toolkit to create SDTM domains in SAS Clinical Data Integration.

Building SDTM Datasets

Now that you have your metadata stored in the SAS Clinical Standards Toolkit, you can build the SDTM domains in Base SAS. In this part of the chapter, we look at changing how you wrote SDTM conversion programs in Chapter 3 so that they can leverage the metadata in the SAS Clinical Standards Toolkit. Except for a few changes, the Base SAS programming is almost entirely the same, so we will dissect only the programming in Chapter 3 that creates the DM/SUPPDM domain.

Base SAS Macros and Tools for SDTM Conversions

In Chapter 3, we introduced the --DTC SDTM date creation macro %make_dtc_date, the --DY SDTM study day macro %make_sdtm_dy, and the codelist format catalog file. For the SAS Clinical Standards Toolkit implementation, you can use those macros and that format catalog exactly as they appeared in Chapter 3. However, you use replacements enabled by the SAS Clinical Standards Toolkit for the %make_empty_dataset macro and the domain-sorting macro %make_sort_order.

Empty Dataset Creator via the SAS Clinical Standards Toolkit

The SAS Clinical Standards Toolkit comes with a SAS macro that creates an empty dataset for you. This dataset is called %cst_createTablesForDataStandard. To create zero record versions of the SDTM datasets defined in the XYZ123 standard, you would run SAS code that looks like this:

```
* initialize the toolkit framework;  ❶
%cst_setStandardProperties(
  _cstStandard=CST-FRAMEWORK,
  _cstSubType=initialize);

* define _ctsGroot macro variable;  ❷
%cstutil_setcstgroot;
%cst_createTablesForDataStandard(
  _cstStandard=CDISC-SDTM,
  _cstStandardVersion=XYZ-SDTM-3.2,
  _cstOutputLibrary=work);            ❸
```

❶ This step initializes the toolkit environment.

❷ This step sets the &_cstgroot macro variable used in step 3.

❸ This step calls the %cst_createTablesForDataStandard macro, creates an empty SAS dataset for every domain dataset found in the XYZ123 standard, and stores those datasets in the SAS WORK library.

Domain Sorter via the SAS Clinical Standards Toolkit

In Chapter 3, there is a SAS macro that sorts the data based on the metadata found in Excel. Here, we modify that macro to sort the data based on the SAS Clinical Standards Toolkit XYZ123 metadata.

```
*----------------------------------------------------------------*;
* make_sort_order.sas creates a global macro variable called
* **SORTSTRING where ** is the name of the dataset that contains
* the metadata specified sort order for a given dataset.
*
* MACRO PARAMETERS:
* dataset = the dataset or domain name
*----------------------------------------------------------------*;
%macro make_sort_order(dataset=);

  ** create **SORTSTRING macro variable;
  %global &dataset.SORTSTRING;
  data _null_;
    set " C:\cstGlobalLibrary\standards\XYZ_program_cdisc-sdtm-3.2-
          1.6\metadata\reference_tables.sas7bdat"; ❶
        where upcase(table) = "&dataset";
        call symputx(compress("&dataset" || "SORTSTRING"),
            left(keys));  ❷

    run;
%mend make_sort_order;
```

When the %make_sort_order macro is executed, the **SORTSTRING global macro variable is created.

❶ Here you use the reference_tables.sas7bdat SAS Clinical Standards Toolkit XYZ123 SDTM domain metadata dataset as the source metadata instead of Excel.

❷ The **SORTSTRING variable is created in this step, which is used in the final dataset sorting when the actual domain is created. The KEYS variables come from the reference_tables.sas7bdat dataset. In this case, we are assuming that you want your final domain sorted by the contents of the KEYS variable as defined in the study metadata. The KEYS variable often describe a unique record for a subject.

Building the Special-Purpose DM and SUPPDM Domains

The following SAS code builds the DM and SUPPDM datasets from our source clinical trial data found in "Appendix A - Source Data Programs." The following program assumes that the source datasets can be found under the libref `source` and that the permanent SAS formats can be found under the libref `library`.

```
*----------------------------------------------------------------*;
* DM.sas creates the SDTM DM and SUPPDM datasets and saves them
* as permanent SAS datasets to the target libref.
*----------------------------------------------------------------*;

**** CREATE EMPTY DM AND SUPPDM DATASETS;   ❶
* initialize the toolkit framework;
%cst_setStandardProperties(
  _cstStandard=CST-FRAMEWORK,
  _cstSubType=initialize);

* define _ctsGroot macro variable;
%cstutil_setcstgroot;

%cst_createTablesForDataStandard(
  _cstStandard=CDISC-SDTM,
  _cstStandardVersion=XYZ-SDTM-3.2,
  _cstOutputLibrary=work
  );


**** GET FIRST AND LAST DOSE DATE FOR RFSTDTC AND RFENDTC;   ❷
proc sort
  data=source.dosing(keep=subject startdt enddt)
  out=dosing;
    by subject startdt enddt;
run;

**** FIRSTDOSE=FIRST DOSING AND LASTDOSE=LAST DOSING;
data dosing;
  set dosing;
    by subject;

    retain firstdose lastdose;

    if first.subject then
      do;
        firstdose = .;
        lastdose = .;
      end;
```

```
        firstdose = min(firstdose,startdt,enddt);
        lastdose = max(lastdose,startdt,enddt);

        if last.subject;
run;

**** GET DEMOGRAPHICS DATA;
proc sort
  data=source.demographic
  out=demographic;
    by subject;
run;

**** MERGE DEMOGRAPHICS AND FIRST DOSE DATE;
data demog_dose;
  merge demographic
        dosing;
    by subject;
run;

**** DERIVE THE MAJORITY OF SDTM DM VARIABLES;          ❸
options missing = ' ';
data dm;
  set dm
    demog_dose(rename=(race=_race));

    studyid = 'XYZ123';
    domain = 'DM';
    usubjid = left(uniqueid);                           ❹
    subjid = put(subject,3.);
    rfstdtc = put(firstdose,yymmdd10.);
    rfendtc = put(lastdose,yymmdd10.);
    rfxstdtc = put(firstdose,yymmdd10.);
    rfxendtc = put(lastdose,yymmdd10.);
    rficdtc = put(icdate,yymmdd10.);
    rfpendtc = put(lastdoc,yymmdd10.);
    dthfl = 'N';
    siteid = substr(subjid,1,1) || "00";
    brthdtc = put(dob,yymmdd10.);
    age = floor ((intck('month',dob,firstdose) -
          (day(firstdose) < day(dob))) / 12);
    if age ne . then
      ageu = 'YEARS';
    sex = put(gender,sex_demographic_gender.);          ❺
    race = put(_race,race_demographic_race.);
    armcd = put(trt,armcd_demographic_trt.);
    arm = put(trt,arm_demographic_trt.);
    actarmcd = put(trt,armcd_demographic_trt.);
    actarm = put(trt,arm_demographic_trt.);
    country = "USA";
  run;


**** DEFINE SUPPDM FOR OTHER RACE;
data suppdm;
  set suppdm
      dm;

    keep studyid rdomain usubjid idvar idvarval qnam qlabel qval qorig
        qeval;

    **** OUTPUT OTHER RACE AS A SUPPDM VALUE;          ❻
    if orace ne '' then
      do;
```

```
              rdomain = 'DM';
              qnam = 'RACEOTH';
              qlabel = 'Race, Other';
              qval = left(orace);
              qorig = 'CRF';
              output;
            end;

          **** OUTPUT RANDOMIZATION DATE AS SUPPDM VALUE;
          if randdt ne . then
            do;
              rdomain = 'DM';
              qnam = 'RANDDTC';
              qlabel = 'Randomization Date';
              qval = left(put(randdt,yymmdd10.));
              qorig = 'CRF';
              output;
            end;
run;


       **** SORT DM ACCORDING TO METADATA AND SAVE PERMANENT DATASET;    ❼
       %make_sort_order(dataset=DM)

       proc sort
         data=dm(keep = studyid domain usubjid subjid rfstdtc rfendtc rfxstdtc
                        rfxendtc rficdtc rfpendtc dthdtc dthfl siteid brthdtc age
                        ageu sex race armcd arm country)
         out=target.dm;
           by &DMSORTSTRING;
       run;


       **** SORT SUPPDM ACCORDING TO METADATA AND SAVE PERMANENT DATASET;
       %make_sort_order(dataset=SUPPDM)

       proc sort
         data=suppdm
         out=target.suppdm;
           by &SUPPDMSORTSTRING;
       run;
```

When the DM.sas program has been run, the DM and SUPPDM SDTM domain datasets are saved to the target libref.

❶   The first step in this program creates the empty SDTM DM and SUPPDM datasets called DM and SUPPDM, based on the SAS Clinical Standards Toolkit metadata.

❷   The SDTM DM variables RFSTDTC and RFENDTC are defined here as the first day of study medication dosing and the last day of dosing, respectively.

❸   This is where the empty dataset DM with the defining variable attributes is set with the source data and the bulk of the SDTM DM variables are defined. Note the RENAME clause in the SET statement, which is required when you have source variables named the same as target variables with conflicting variable attributes.

❹   In these examples, you see that there is a source variable called UNIQUEID in the source datasets, which is supposed to uniquely identify a patient within and across studies. In most clinical databases, a truly unique patient identifier UNIQUEID does not exist, and in practice many trials are submitted where USUBJID is set equal to SUBJID.

❺ For the SEX, RACE, ARM, and ARMCD variables, the associated formats were created in the make_codelist_formats.sas program in the previous section. In this example, ARM is the same as ACTARM, which is fairly simplistic. It in no way means that your study will have all patients receiving assigned treatment.

❻ The supplemental qualifiers created here in SUPPDM are for other race and randomization date. This chunk of code creates those qualifier records. However, you typically have to define IDVAR and IDVARVAL. IDVAR defines the SDTM parent domain variable. IDVARVAL defines the value of that variable that can be used to join or merge these supplemental qualifier variables back onto the parent domain. Often, IDVAR=--SEQ and IDVARVAL is the sequence number because --SEQ tends to be the most exacting identifier that you can have in the SDTM because it identifies a single record. You should define IDVAR to be the highest-level identifier that you can in order to be clear about the qualifying relationship to the parent domain.

❼ At this point in the program, we call the %make_sort_order macro twice to get the DMSORTSTRING and SUPPDMSORTSTRING macro variables defined. Then we use that to sort and save our final DM and SUPPDM domain datasets.

The remaining domains created with Base SAS in Chapter 3 can be modified like this to use the SAS Clinical Standards Toolkit metadata instead of Microsoft Excel. The key changes involve the empty dataset creation and the final sort order definition so that those are based on the SAS Clinical Standards Toolkit instead of Excel. However, the remaining code from Chapter 3 is essentially untouched here in Chapter 4.

Building Define.xml

Now that you have built your SDTM datasets based on the XYZ123 custom SDTM standard in the SAS Clinical Standards Toolkit, you can build your define.xml file using the SAS Clinical Standards Toolkit. The first thing to do is to create an area to store your metadata in order to build your define file. For this example, the C:\Users\yourid\Desktop\SAS_BOOK_cdisc_2nd_edition\CST\CST_make_define folder was created for this work as the "root" level for the define file work. Under this root level, you will want to create additional folders called "metadata," "results," "sourcedata," "sourcexml," and "stylesheet." These folders will contain the following content:

Folder	Contents
"root"	This is where you will store your CST programs
\metadata	This is where you will create SAS datasets source_study, source_tables, source_values, source_codelists, and source_documents that drive the define.xml creation process.
\results	This is where create_sasdefine_from_source.sas and create_definexml.sas will write informational run log datasets that you can review.
\sourcedata	This is where create_sasdefine_from_source.sas writes datasets for create_definexml.sas to use.
\sourcexml	This is where created_definexml.sas will save your final define.xml and associated XSL file.

Folder	Contents
\stylesheet	This is the template XSL style sheet that create_definexml.sas uses and places into \sourcexml.

Keep in mind that the CST is designed to be run in UTF-8 encoding, so you may need to switch your SAS session encoding from Latin to UTF-8 before running any programs below.

Create the Study Level Metadata Dataset

You need to first define a study level metadata dataset called C:\Users\yourid\Desktop\SAS_BOOK_cdisc_2nd_edition\CST\CST_make_define\metadata\sour ce_study.sas7bdat. This is manually created study-level metadata data that ends up in the define file header. For this example, that dataset looks like this:

	standard	standardversion	FORMALSTANDARDNAME	FORMALSTANDARDVERSION	PROTOCOLNAME	SASREF	STUDYDESCRIPTION	STUDYNAME	STUDYVERSION
1	CDISC-SDTM	XYZ-SDTM-3.2	SDTM-IG	3.2	XYZ123	SRCDATA	A PHASE IIB, DOUBLE-BLIND, MULTI-CENTER, PLACEBO CONTROLLED, PARALLEL GROUP TRIAL OF ANALGEZIA HCL FOR THE TREATMENT OF CHRONIC PAIN	XYZ123	study1

Create the Table and Column-Level Metadata Datasets

Now you need to create the table and column-level metadata datasets, which you can get from your CST defined standard. Copy C:\cstGlobalLibrary\standards\XYZ_program_cdisc-sdtm-3.2-1.6\metadata reference_columns and reference_tables SAS datasets for the XYZ123 standard to the define file metadata folder of C:\Users\yourid\Desktop\SAS_BOOK_cdisc_2nd_edition\CST\CST_make_define\metadata source_columns.sas7bdat and source_tables.sas7bdat respectively. Subset those datasets just for the domains and variables found in the XYZ123 SDTM datasets.

The define file generation process requires a few additional columns that are not found in the reference_tables and reference_columns datasets that you do need to add for the define file generation to work properly. When you copy reference_tables to source_tables, you need to add columns StudyVersion, Domain, Order, and DomainDescription. When copying reference_columns to source_columns, you need to add columns StudyVersion, SignificantDigits, and OriginDescription.

Create the Value-Level Metadata Dataset

Value-level metadata is essentially row-level metadata that enables you to further distinguish more precise metadata characteristics of a column value. Think of laboratory data and how the results can have varying types (for example, character or numeric) or varying degrees of numeric precision (that is, significant digits). For the SDTM, some level of value-level metadata is expected, but the precise quantity needed is left undefined by CDISC. For this example, the only value-level metadata that we specify is around the lab data. It is as follows with the null value variables of ALGORITHM, COMMENT, VALUE, and WHERECLAUSECOMMENT columns hidden. Here is a snapshot of the source_values dataset:

	standard	standardversion	COLUMN	CORE	DISPLAYFORMAT	LABEL	LENGTH	ORDER	ORIGIN	ROLE
1	CDISC-SDTM	XYZ-SDTM-3.2	LBSTRESN	Yes	4.2	Result or Finding in Original Units	8	1	eDT	ResultQualifier
2	CDISC-SDTM	XYZ-SDTM-3.2	LBSTRESN	Yes	3.0	Result or Finding in Original Units	8	2	eDT	ResultQualifier
3	CDISC-SDTM	XYZ-SDTM-3.2	LBSTRESN	Yes	3.0	Result or Finding in Original Units	8	3	eDT	ResultQualifier
4	CDISC-SDTM	XYZ-SDTM-3.2	LBSTRESN	Yes	3.0	Result or Finding in Original Units	8	4	eDT	ResultQualifier
5	CDISC-SDTM	XYZ-SDTM-3.2	LBSTRESN	Yes	4.2	Result or Finding in Original Units	8	5	eDT	ResultQualifier
6	CDISC-SDTM	XYZ-SDTM-3.2	LBSTRESN	Yes	4.2	Result or Finding in Original Units	8	6	eDT	ResultQualifier
7	CDISC-SDTM	XYZ-SDTM-3.2	LBSTRESN	Yes	3.0	Result or Finding in Original Units	8	7	eDT	ResultQualifier
8	CDISC-SDTM	XYZ-SDTM-3.2	LBSTRESN	Yes	4.2	Result or Finding in Original Units	8	8	eDT	ResultQualifier
9	CDISC-SDTM	XYZ-SDTM-3.2	LBSTRESC	Yes		Result or Finding in Original Units	200	9	eDT	ResultQualifier
10	CDISC-SDTM	XYZ-SDTM-3.2	LBSTRESN	Yes	3.0	Result or Finding in Original Units	8	10	eDT	ResultQualifier
11	CDISC-SDTM	XYZ-SDTM-3.2	LBSTRESN	Yes	5.2	Result or Finding in Original Units	8	11	eDT	ResultQualifier
12	CDISC-SDTM	XYZ-SDTM-3.2	LBSTRESN	Yes	4.2	Result or Finding in Original Units	8	12	eDT	ResultQualifier

	SASREF	SIGNIFICANTDIGITS	STUDYVERSION	TABLE	TYPE	WHERECLAUSE	XMLCODELIST	XMLDATATYPE
1	SRCDATA	2	study1	LB	N	LBTESTCD EQ "ALB"		float
2	SRCDATA	.	study1	LB	N	LBTESTCD EQ "ALP"		integer
3	SRCDATA	.	study1	LB	N	LBTESTCD EQ "ALT"		integer
4	SRCDATA	.	study1	LB	N	LBTESTCD EQ "AST"		integer
5	SRCDATA	2	study1	LB	N	LBTESTCD EQ "BILDUR"		float
6	SRCDATA	2	study1	LB	N	LBTESTCD EQ "BILI"		float
7	SRCDATA	.	study1	LB	N	LBTESTCD EQ "GGT"		integer
8	SRCDATA	2	study1	LB	N	(LBCAT EQ "CHEMISTRTY") AND (LBTESTCD EQ "GLUC")		float
9	SRCDATA	.	study1	LB	C	(LBCAT EQ "URINALYSIS") AND (LBTESTCD EQ "GLUC")	URINGLUC	text
10	SRCDATA	.	study1	LB	N	LBTESTCD EQ "HCT"		integer
11	SRCDATA	2	study1	LB	N	LBTESTCD EQ "HGB"		float
12	SRCDATA	2	study1	LB	N	LBTESTCD EQ "PROT"		float

Create the Codelists Metadata Dataset

For every column in the source_columns or source_values dataset that has an XMLCODELIST defined, you need to create a codelist in the source_codelist dataset. Here is what that dataset looks like with the null value variables of CODEDVALUENUM, CODELISTDESCRIPTION, EXTENDEDVALUE, HREF, and REF hidden. Also note that this is only a partial selection of all of the rows in the source_codelists dataset.

	standard	standardversion	CODEDVALUECHAR	CODEDVALUENCICODE	CODELIST	CODELISTDATATYPE	CODELISTNAME	CODELISTNCICODE
1	CDISC-SDTM	XYZ-SDTM-3.2	MILD	C41338	AESEV	text	Severity/Intensity Scale for Adverse Events	C66769
2	CDISC-SDTM	XYZ-SDTM-3.2	MODERATE	C41339	AESEV	text	Severity/Intensity Scale for Adverse Events	C66769
3	CDISC-SDTM	XYZ-SDTM-3.2	SEVERE	C41340	AESEV	text	Severity/Intensity Scale for Adverse Events	C66769
8	CDISC-SDTM	XYZ-SDTM-3.2	M	C20197	SEX	text	Sex	C66731
9	CDISC-SDTM	XYZ-SDTM-3.2	F	C16576	SEX	text	Sex	C66731
10	CDISC-SDTM	XYZ-SDTM-3.2	WHITE	C41261	RACE	text	Race	C74457
11	CDISC-SDTM	XYZ-SDTM-3.2	BLACK OR AFRICAN AMERICAN	C16352	RACE	text	Race	C74457
12	CDISC-SDTM	XYZ-SDTM-3.2	ASIAN	C41260	RACE	text	Race	C74457
26	CDISC-SDTM	XYZ-SDTM-3.2	USA	C17234	COUNTRY	text	Country	C66786
36	CDISC-SDTM	XYZ-SDTM-3.2	Baseline		VISIT	text	Visit	
37	CDISC-SDTM	XYZ-SDTM-3.2	3 Months		VISIT	text	Visit	
38	CDISC-SDTM	XYZ-SDTM-3.2	6 Months		VISIT	text	Visit	
70	CDISC-SDTM	XYZ-SDTM-3.2			AEDECOD	text	AE Decode	
71	CDISC-SDTM	XYZ-SDTM-3.2			AEBODSYS	text	AE Body System	

	DECODELANGUAGE	DECODETEXT	DICTIONARY	ORDERNUMBER	RANK	SASFORMATNAME	SASREF	STUDYVERSION	VERSION
1	en	Grade 1; 1		1	1	$AESEV	SRCDATA	study1	
2	en	Grade 2; 2		2	2	$AESEV	SRCDATA	study1	
3	en	Grade 3; 3		3	3	$AESEV	SRCDATA	study1	
8	en	Male		1		$SEX	SRCDATA	study1	
9	en	Female		2		$SEX	SRCDATA	study1	
10	en	White		1		$RACE	SRCDATA	study1	
11	en	African American		2		$RACE	SRCDATA	study1	
12	en	Asian		3		$RACE	SRCDATA	study1	
26	en	UNITED STATES		1		$COUNTRY	SRCDATA	study1	
36	en			1		$VISIT	SRCDATA	study1	
37	en			2		$VISIT	SRCDATA	study1	
38	en			3		$VISIT	SRCDATA	study1	
70	en		MedDRA	1		$AEDECOD	SRCDATA	study1	7.0
71	en		MedDRA	1		$AEBODSYS	SRCDATA	study1	7.0

Create the Document Reference Metadata Dataset

Hyperlinks to external documents from the define.xml file is are handled by the CST in a document reference dataset called source_documents. You need to create a row in this dataset for each document reference that you wish.

	standard	standardversion	COLUMN	DOCTYPE	HREF	PDFPAGEREFS	PDFPAGEREFTYPE	SASREF	STUDYVERSION	TABLE	TITLE	WHERECLAUSE
1	CDISC-SDTM	XYZ-SDTM-3.2	AFTFRM	CRF	blankcrf.pdf	6	PhysicalRef		study1	AE	Annotated CRF	
2	CDISC-SDTM	XYZ-SDTM-3.2	AESEV	CRF	blankcrf.pdf	6	PhysicalRef		study1	AE	Annotated CRF	
3	CDISC-SDTM	XYZ-SDTM-3.2	AESER	CRF	blankcrf.pdf	6	PhysicalRef		study1	AE	Annotated CRF	
4	CDISC-SDTM	XYZ-SDTM-3.2	AEACN	CRF	blankcrf.pdf	6	PhysicalRef		study1	AE	Annotated CRF	
5	CDISC-SDTM	XYZ-SDTM-3.2	AEREL	CRF	blankcrf.pdf	6	PhysicalRef		study1	AE	Annotated CRF	
6	CDISC-SDTM	XYZ-SDTM-3.2	AESTDTC	CRF	blankcrf.pdf	6	PhysicalRef		study1	AE	Annotated CRF	
7	CDISC-SDTM	XYZ-SDTM-3.2	AEENDTC	CRF	blankcrf.pdf	6	PhysicalRef		study1	AE	Annotated CRF	
8	CDISC-SDTM	XYZ-SDTM-3.2	SUBJID	CRF	blankcrf.pdf	1	PhysicalRef		study1	DM	Annotated CRF	
9	CDISC-SDTM	XYZ-SDTM-3.2	RFSTDTC	CRF	blankcrf.pdf	1	PhysicalRef		study1	DM	Annotated CRF	
10	CDISC-SDTM	XYZ-SDTM-3.2	SITEID	CRF	blankcrf.pdf	1	PhysicalRef		study1	DM	Annotated CRF	
11	CDISC-SDTM	XYZ-SDTM-3.2	BRTHDTC	CRF	blankcrf.pdf	1	PhysicalRef		study1	DM	Annotated CRF	
12	CDISC-SDTM	XYZ-SDTM-3.2	SEX	CRF	blankcrf.pdf	1	PhysicalRef		study1	DM	Annotated CRF	
13	CDISC-SDTM	XYZ-SDTM-3.2	RACE	CRF	blankcrf.pdf	1	PhysicalRef		study1	DM	Annotated CRF	
14	CDISC-SDTM	XYZ-SDTM-3.2	EXDOSE	CRF	blankcrf.pdf	4	PhysicalRef		study1	EX	Annotated CRF	
15	CDISC-SDTM	XYZ-SDTM-3.2	EXDOSTOT	CRF	blankcrf.pdf	4	PhysicalRef		study1	EX	Annotated CRF	
16	CDISC-SDTM	XYZ-SDTM-3.2	EXSTDTC	CRF	blankcrf.pdf	4	PhysicalRef		study1	EX	Annotated CRF	
17	CDISC-SDTM	XYZ-SDTM-3.2	QVAL	CRF	blankcrf.pdf	1	PhysicalRef		study1	SUPPDM	Annotated CRF	
18	CDISC-SDTM	XYZ-SDTM-3.2	XPORRES	CRF	blankcrf.pdf	5	PhysicalRef		study1	XP	Annotated CRF	
19	CDISC-SDTM	XYZ-SDTM-3.2	VISIT	CRF	blankcrf.pdf	5	PhysicalRef		study1	XP	Annotated CRF	
20	CDISC-SDTM	XYZ-SDTM-3.2	XPDTC	CRF	blankcrf.pdf	5	PhysicalRef		study1	XP	Annotated CRF	

Create Define File Generation Programs

Now that the SAS Clinical Standards Toolkit define file metadata has been created, you can create the define file generation programs. There are two steps to generating the define file. First you must run a program that creates the source datasets that feed the define file. Then you must run a program that uses those datasets to create the define.xml file itself.

To create the program that creates the datasets that are used to write the define file, copy create_sasdefine_from_source.sas from: C:\cstSampleLibrary\cdisc-definexml-2.0.0-1.6\programs to C:\Users\yourid\Desktop\SAS_BOOK_cdisc_2nd_edition\CST\CST_make_define. After a few edits, the final create_sasdefine_from_source.sas program looks like this:

```
*****************************************************************************;
* create_sasdefine_from_source.sas                                        *;
*                                                                         *;
* Sample driver program to perform a primary Toolkit action, in this case, *;
* creating the Define-XML data sets from the CDISC standard model (SDTM,   *;
* ADaM or SEND source metadata).                                          *;
* A call to a standard-specific Define-XML macro is required later in this  *;
* code.                                                                    *;
*                                                                         *;
* Assumptions:                                                            *;
*        The SASReferences file must exist, and must be identified in the   *;
```

```
*           call to cstutil_processsetup if it is not work.sasreferences.     *;
*                                                                             *;
* CSTversion   1.6                                                            *;
*                                                                             *;
*   The following statements may require information from the user            *;
******************************************************************************  ******;

%let _cstStandard=CDISC-DEFINE-XML;
%let _cstStandardVersion=2.0.0;      * <----- User sets the Define-XML version *;

%let _cstTrgStandard=CDISC-SDTM;    * <----- User sets to standard of the
study*;❶
%let _cstTrgStandardVersion=XYZ-SDTM-3.2; * <----- User sets to standard
version      *;

%*let _cstTrgStandard=CDISC-ADAM;    * <----- User sets to standard of the study
*;
%*let _cstTrgStandardVersion=2.1;   * <----- User sets to standard version     *;

%* Subfolder with the SAS Source Metadata data sets;
%let _cstSrcMetaDataFolder=%lowcase(&_cstTrgStandard)-
&_cstTrgStandardVersion/metadata;


******************************************************************************;
* The following code sets (at a minimum) the studyrootpath and               *;
* studyoutputpath. These are used to make the driver programs portable across*;
* platforms and allow the code to be run with minimal modification. These   *;
* macro variables by default point to locations within the cstSampleLibrary, *;
* set during install but modifiable thereafter.  The cstSampleLibrary is      *;
* assumed to allow write operations by this driver module.                   *;
******************************************************************************;

%cst_setStandardProperties(_cstStandard=CST-FRAMEWORK, _cstSubType=initialize);
%cstutil_setcstsroot;
data _null_;
call symput
('studyRootPath',cats("C:\Users\yourid\Desktop\SAS_BOOK_cdisc_2nd_edition\CST\C
ST_make_define"));    ❷
call symput
('studyOutputPath',cats("C:\Users\yourid\Desktop\SAS_BOOK_cdisc_2nd_edition\CST
\CST_make_define"));  ❷
run;
%let workPath=%sysfunc(pathname(work));




******************************************************************************;
* One strategy to defining the required library and file metadata for a CST  *;
* process is to optionally build SASReferences in the WORK library.  An       *;
* example of how to do this follows.                                         *;
*                                                                            *;
* The call to cstutil_processsetup below tells CST how SASReferences will be *;
* provided and referenced. If SASReferences is built in work, the call to    *;
* cstutil_processsetup may, assuming all defaults, be as simple as:          *;
*         %cstutil_processsetup(_cstStandard=CDISC-SDTM)                      *;
******************************************************************************;

%let _cstSetupSrc=SASREFERENCES;

%cst_createdsfromtemplate(_cstStandard=CST-FRAMEWORK,
_cstType=control,_cstSubType=reference, _cstOutputDS=work.sasreferences);
```

```
proc sql;                                    ❸
  insert into work.sasreferences
  values ("CST-FRAMEWORK"      "1.2"                        "messages"        ""
"messages" "libref"  "input"  "dataset"  "N"  "" ""                       1
""                      "")
  values ("&_cstStandard"       "&_cstStandardVersion"       "messages"       ""
"crtmsg"   "libref"  "input"  "dataset"  "N"  "" ""                      2
""                      "")
  values ("&_cstStandard"       "&_cstStandardVersion"       "autocall"       ""
"auto1"    "fileref" "input"  "folder"   "N"  "" ""                      1
""                      "")
  values ("&_cstStandard"       "&_cstStandardVersion"       "properties"
"initialize" "inprop"  "fileref" "input"  "file"    "N"  "" ""
1  ""                      "")
  values ("&_cstStandard"       "&_cstStandardVersion"       "results"
"results"     "results"  "libref"  "output" "dataset"  "Y"  ""  ""
"&studyOutputPath/results"  .  "sourcetodefine_results.sas7bdat" "")
  values ("&_cstStandard"       "&_cstStandardVersion"       "sourcedata"     ""
"srcdata"  "libref"  "output" "folder"   "Y"  ""  "&studyOutputPath/sourcedata"
.  ""     "")
  values ("&_cstTrgStandard"  "&_cstTrgStandardVersion" "sourcemetadata"
"study"       "sampdata" "libref"  "input"  "dataset"  "N"  ""
"&studyRootPath/metadata"  .  "source_study"            "")
  values ("&_cstTrgStandard"  "&_cstTrgStandardVersion" "sourcemetadata"
"table"       "sampdata" "libref"  "input"  "dataset"  "N"  ""
"&studyRootPath/metadata"  .  "source_tables"           "")
  values ("&_cstTrgStandard"  "&_cstTrgStandardVersion" "sourcemetadata"
"column"      "sampdata" "libref"  "input"  "dataset"  "N"  ""
"&studyRootPath/metadata"  .  "source_columns"          "")
  values ("&_cstTrgStandard"  "&_cstTrgStandardVersion" "sourcemetadata"
"codelist"    "sampdata" "libref"  "input"  "dataset"  "N"  ""
"&studyRootPath/metadata"  .  "source_codelists"        "")
  values ("&_cstTrgStandard"  "&_cstTrgStandardVersion" "sourcemetadata"
"value"       "sampdata" "libref"  "input"  "dataset"  "N"  ""
"&studyRootPath/metadata"  .  "source_values"           "")
  values ("&_cstTrgStandard"  "&_cstTrgStandardVersion" "sourcemetadata"
"document"    "sampdata" "libref"  "input"  "dataset"  "N"  ""
"&studyRootPath/metadata"  .  "source_documents"        "")
;
quit;


*****************************************************************;
* Debugging aid:  set _cstDebug=1                             *;
* Note value may be reset in call to cstutil_processsetup     *;
*  based on property settings.  It can be reset at any        *;
*  point in the process.                                      *;
*****************************************************************;
%let _cstDebug=0;
data _null_;
  _cstDebug = input(symget('_cstDebug'),8.);
  if _cstDebug then
    call execute("options &_cstDebugOptions;");
  else
    call execute(("%sysfunc(tranwrd(options %cmpres(&_cstDebugOptions), %str( ),
%str( no)));"));
run;


***********************************************************************************;
* Clinical Standards Toolkit utilizes autocall macro libraries to contain       *;
* and reference standard-specific code libraries. Once the autocall path is     *;
* set and one or more macros have been used within any given autocall           *;
* library deallocation or reallocation of the autocall fileref cannot occur     *;
* unless the autocall path is first reset to exclude the specific fileref.      *;
*                                                                               *;
```

```
* This becomes a problem only with repeated calls to %cstutil_processsetup() *;
* or %cstutil_allocatesasreferences within the same sas session.  Doing so,  *;
* without submitting code similar to the code below may produce SAS errors    *;
* such as:                                                                    *;
*   ERROR - At least one file associated with fileref AUTO1 is still in use.  *;
*   ERROR - Error in the FILENAME statement.                                  *;
*                                                                             *;
* If you call %cstutil_processsetup() or %cstutil_allocatesasreferences more  *;
* than once within the same sas session, typically using %let                 *;
* _cstReallocateSASRefs=1 to tell CST to attempt reallocation, use of the     *;
* following code is recommended between each code submission.                 *;
*                                                                             *;
* Use of the following code is NOT needed to run this driver module           *;
* initially.                                                                  *;
*******************************************************************************;

%*let _cstReallocateSASRefs=1;
%*include "&_cstGRoot/standards/cst-framework-
&_cstVersion/programs/resetautocallpath.sas";


*******************************************************************************;
* The following macro (cstutil_processsetup) utilizes the following          *;
* parameters:                                                                 *;
*                                                                             *;
* _cstSASReferencesSource - Setup should be based upon what initial source?   *;
*   Values: SASREFERENCES (default) or RESULTS data set. If RESULTS:          *;
*      (1) no other parameters are required and setup responsibility is       *;
*               passed to the cstutil_reportsetup macro                       *;
*      (2) the results data set name must be passed to cstutil_reportsetup as *;
*               libref.memname                                                *;
*                                                                             *;
* _cstSASReferencesLocation - The path (folder location) of the              *;
*                             sasreferences data set (default is the path to  *;
*                             the WORK library)                               *;
*                                                                             *;
* _cstSASReferencesName - The name of the sasreferences data set             *;
*                         (default is sasreferences)                          *;
*******************************************************************************;

%cstutil_processsetup();

*******************************************************************************;
* Run the standard-specific Define-XML macros.                               *;
*******************************************************************************
;

%define_sourcetodefine(
  _cstOutLib=srcdata,
  _cstSourceStudy=sampdata.source_study,
  _cstSourceTables=sampdata.source_tables,
  _cstSourceColumns=sampdata.source_columns,
  _cstSourceCodeLists=sampdata.source_codelists,
  _cstSourceDocuments=sampdata.source_documents,
  _cstSourceValues=sampdata.source_values,
  _cstFullModel=N,
  _cstLang=en
  );

*******************************************************************************;
* Clean-up the CST process files, macro variables and macros.                *;
*******************************************************************************;
* Delete sasreferences if created above  *;
proc datasets lib=work nolist;
```

```
    delete sasreferences / memtype=data;
quit;

%*cstutil_cleanupcstsession(
     _cstClearCompiledMacros=0
    ,_cstClearLibRefs=1
    ,_cstResetSASAutos=1
    ,_cstResetFmtSearch=0
    ,_cstResetSASOptions=0
    ,_cstDeleteFiles=1
    ,_cstDeleteGlobalMacroVars=0);
```

❶ You need to define the _cstTrgStandard and _cstTrgStandardVersion macro variables to point at your standard for the study.

❷ The studyRootpath and studyOutputPath need to point to your define file work area.

❸ This is where you need to define your sasreferences dataset. It defines where the CST can find things. If you set up the folder structure as indicated at the start of this section, then this section can be left as it is.

Any findings from running this program would be written to the \results\sourcetodefine_results.sas7bdat SAS dataset. After create_sasdefine_from_source.sas has been run, it creates the source datasets needed for the SAS Clinical Standards Toolkit to create the define.xml file. Those datasets are stored at C:\Users\yourid\Desktop\SAS_BOOK_cdisc_2nd_edition\CST\CST_make_define\sourcedata. The datasets look like this:

Now you need to run the SAS program that takes those SAS datasets and writes the actual define file. Copy create_definexml.sas from C:\cstSampleLibrary\cdisc-definexml-2.0.0-1.6\programs to C:\Users\yourid\Desktop\SAS_BOOK_cdisc_2nd_edition\CST\CST_make_define. The final create_definexml.sas program that writes the define file looks like this:

```
***************************************************************************;
* create_definexml.sas                                                   *;
*                                                                        *;
```

```
* Sample driver program to create a CDISC-DEFINE-XML V2.0.0 define.xml file  *;
*                                                                            *;
* Assumptions:                                                               *;
*          The SASReferences file must exist, and must be identified in the  *;
*          call to cstutil_processsetup if it is not work.sasreferences.     *;
*                                                                            *;
* CSTversion  1.6                                                            *;
*                                                                            *;
*  The following statements may require information from the user            *;
*****************************************************************************;

%let _cstStandard=CDISC-DEFINE-XML;
%let _cstStandardVersion=2.0.0;

%let _cstTrgStandard=CDISC-SDTM;          * <----- User sets to standard
*;❶
%let _cstTrgStandardVersion=XYZ-SDTM-3.2; * <----- User sets to standard
version*;
%let _cstDefineFile=define.xml;

%*let _cstTrgStandard=CDISC-ADAM;       * <----- User sets to standard         *;
%*let _cstTrgStandardVersion=2.1;       * <----- User sets to standard version *;
%*let _cstDefineFile=define.xml;

%* Subfolder with the SAS Define-XML data sets;
%let _cstSrcDataFolder=%lowcase(&_cstTrgStandard)-&_cstTrgStandardVersion;

*****************************************************************************;
* The following data step sets (at a minimum) the studyrootpath and         *;
* studyoutputpath. These are used to make the driver programs portable       *;
* across platforms and allow the code to be run with minimal modification.   *;
* These nacro variables by default point to locations within the            *;
* cstSampleLibrary, set during install but modifiable thereafter. The        *;
* cstSampleLibrary is assumed to allow write operations by this driver       *;
* module.                                                                    *;
*************** *************************************************************;

%cst_setStandardProperties(_cstStandard=CST-FRAMEWORK,_cstSubType=initialize);
%cstutil_setcstsroot;
data _null_;
call symput
('studyRootPath',cats("C:\Users\yourid\Desktop\SAS_BOOK_cdisc_2nd_edition\CST\C
ST_make_define"));  ❷
call symput
('studyOutputPath',cats("C:\Users\yourid\Desktop\SAS_BOOK_cdisc_2nd_edition\CST
\CST_make_define"));❷
run;
%let workPath=%sysfunc(pathname(work));


*****************************************************************************;
* One strategy to defining the required library and file metadata for a CST  *;
* process is to optionally build SASReferences in the WORK library.  An      *;
* example of how to do this follows.                                         *;
*                                                                            *;
* The call to cstutil_processsetup below tells CST how SASReferences will be *;
* provided and referenced. If SASReferences is built in work, the call to    *;
* cstutil_ processsetup may, assuming all defaults, be as simple as:          *;
*         %cstutil_processsetup(_cstStandard=CDISC-SDTM)                      *;
*****************************************************************************;

%let _cstSetupSrc=SASREFERENCES;
```

```
%cst_createdsfromtemplate(_cstStandard=CST-FRAMEWORK,
_cstType=control,_cstSubType=reference, _cstOutputDS=work.sasreferences);

proc sql;                                       ❸
  insert into work.sasreferences
  values ("CST-FRAMEWORK"  "1.2"                         "messages"      ""
"messages" "libref"  "input"  "dataset"  "N"   ""   ""
1 ""                        "")
  values ("&_cstStandard"  "&_cstStandardVersion"  "messages"       ""
"crtmsg"    "libref"  "input"  "dataset"  "N"   ""   ""
2 ""                        "")
  values ("&_cstStandard"  "&_cstStandardVersion"  "autocall"       ""
"auto1"    "fileref" "input"  "folder"   "N"   ""   ""
1 ""                        "")
  values ("&_cstStandard"  "&_cstStandardVersion"  "control"        "reference"
"control"  "libref"  "both"   "dataset"  "Y"   ""   "&workpath"
. "sasreferences"             "")
  values ("&_cstStandard"  "&_cstStandardVersion"  "results"        "results"
"results"  "libref"  "output" "dataset"  "Y"   ""   "&studyOutputPath/results"
. "write_results.sas7bdat" "")
  values ("&_cstStandard"  "&_cstStandardVersion"  "sourcedata"     ""
"srcdata"  "libref"  "input"  "folder"   "N"   ""   "&studyRootPath/sourcedata"
. ""                        "")
  values ("&_cstStandard"  "&_cstStandardVersion"  "externalxml"  "xml"
"extxml"   "fileref" "output" "file"     "Y"   ""   "&studyOutputPath/sourcexml"
. "&_cstDefineFile"          "")
  values ("&_cstStandard"  "&_cstStandardVersion"  "referencexml" "stylesheet"
"xslt01"   "fileref" "output" "file"     "Y"   ""   "&studyOutputPath/stylesheet"
. "define2-0-0.xsl"          "")
  values ("&_cstStandard"  "&_cstStandardVersion"  "properties"     "initialize"
"inprop"   "fileref" "input"  "file"     "N"   ""   ""
1 ""                        "")
  ;
quit;


***************************************************************;
* Debugging aid:  set _cstDebug=1                        *;
* Note value may be reset in call to cstutil_processsetup *;
*  based on property settings.  It can be reset at any    *;
*  point in the process.                                  *;
***************************************************************;
%let _cstDebug=0;
data _null_;
  _cstDebug = input(symget('_cstDebug'),8.);
  if _cstDebug then
    call execute("options &_cstDebugOptions;");
  else
    call execute(("%sysfunc(tranwrd(options %cmpres(&_cstDebugOptions), %str(
), %str( no)));"));
run;


*********************************************************************************;
* Clinical Standards Toolkit utilizes autocall macro libraries to contain and*;
* reference standard-specific code libraries.  Once the autocall path is set *;
* and one or more macros have been used within any given autocall library,   *;
* deallocation or reallocation of the autocall fileref cannot occur unless   *;
* the autocall path is first reset to exclude the specific fileref.          *;
*                                                                            *;
* This becomes a problem only with repeated calls to %cstutil_processsetup() *;
* or %cstutil_allocatesasreferences within the same sas session.  Doing so,  *;
* without submitting code similar to the code below may produce SAS errors   *;
* such as:                                                                   *;
*   ERROR - At least one file associated with fileref AUTO1 is still in use. *;
```

```
*       ERROR - Error in the FILENAME statement.               *;
*                                                              *;
* If you call %cstutil_processsetup() or %cstutil_allocatesasreferences more *;
* than once within the same sas session, typically using %let  *;
* _cstReallocateSASRefs=1 to tell CST to attempt reallocation, use of the *;
* following code is recommended between each code submission.   *;
*                                                              *;
* Use of the following code is NOT needed to run this driver module *;
* initially.                                                    *;
****************************************************************************;

%*let _cstReallocateSASRefs=1;
%*include "&_cstGRoot/standards/cst-framework-
&_cstVersion/programs/resetautocallpath.sas";


****************************************************************************;
* The following macro (cstutil_processsetup) utilizes the following   *;
* parameters:                                                          *;
*                                                                      *;
* _cstSASReferencesSource - Setup should be based upon what initial source? *;
*    Values: SASREFERENCES (default) or RESULTS data set. If RESULTS: *;
*    (1) no other parameters are required and setup responsibility is *;
*           passed to the cstutil_reportsetup macro                   *;
*    (2) the results data set name must be passed to cstutil_reportsetup as *;
*           libref.memname                                            *;
*                                                                      *;
* _cstSASReferencesLocation - The path (folder location) of the       *;
*                             sasreferences data set (default is      *;
*                             the path to the WORK library)           *;
*                                                                      *;
* _cstSASReferencesName - The name of the sasreferences data set      *;
*                         (default is sasreferences)                  *;
************************************************************ ****************;

%cstutil_processsetup();


****************************************************************************
;
* Run the standard-specific Define-XML macros.
*;
****************************************************************************
;
%define_write(
  _cstCreateDisplayStyleSheet=1,
  _cstHeaderComment=%str( Produced from SAS data using the SAS Clinical
Standards Toolkit &_cstVersion )
  );

****************************************************************************;
* Run the cross-standard schema validation macro.                    *;
* Running cstutilxmlvalidate is not required.  The define_read macro will *;
* attempt to import an invalid define xml file. However, importing an  *;
* invalid define xml file may result in an incomplete import.          *;
*                                                                      *;
* cstutilxmlvalidate parameters (all optional):                        *;
*   _cstSASReferences:  The SASReferences data set provides the location *:
*           of the to-be-validate XML file associated with a registered *;
*           standard and standardversion (default:  &_cstSASRefs).    *;
*   _cstLogLevel:  Identifies the level of error reporting.            *;
*           Valid values: Info (default) Warning, Error, Fatal Error   *;
*   _cstCallingPgm:  The name of the driver module calling this macro  *;
****************************************************************************;
```

```
%cstutilxmlvalidate();

****************** ************************************************************;
* Clean-up the CST process files, macro variables and macros.              *;
****************** ************************************************** **************;
* Delete sasreferences if created above  *;
proc datasets lib=work nolist;
  delete sasreferences / memtype=data;
quit;

%*cstutil_cleanupcstsession(
    _cstClearCompiledMacros=0
  , _cstClearLibRefs=1
  , _cstResetSASAutos=1
  , _cstResetFmtSearch=0
  , _cstResetSASOptions=0
  , _cstDeleteFiles=1
  , _cstDeleteGlobalMacroVars=0);
```

❶ You need to define the _cstTrgStandard and _cstTrgStandardVersion macro variables to point at your standard for the study.

❷ The studyRootpath and studyOutputPath need to point to your define file work area.

❸ This is where you need to define your sasreferences dataset. It defines where the CST can find things. If you set up the folder structure as indicated at the start of this section, then this section can be left as it is.

The %define_write macro actually writes the define.xml file where the _cstCreateDisplayStyleSheet=1 macro variable setting creates the default CDISC supplied style sheet. The %cstutilxmlvalidate macro ensures that the define.xml file is a valid XML file. Any findings from the validation or define file generation would be written to the \results\write_results.sas7bdat SAS dataset. The start of the resulting define.xml file looks like this:

Chapter Summary

In this chapter, you used the SAS Clinical Standards Toolkit to define and register a new XYZ123 compound customized SDTM standard. You then used that XYZ123 standard to build SDTM domains within Base SAS code. This chapter shows you how you can use the SAS Clinical Standards Toolkit to house your CDISC SDTM metadata instead of Microsoft Excel (as seen in Chapter 2). The Base SAS SDTM conversion code from Chapter 3 was then updated to use the SAS Clinical Standards Toolkit metadata for conversion purposes. Finally, we used the SAS Clinical Standards Toolkit to create the define.xml file.

Chapter 5: Implementing the CDISC SDTM with SAS Clinical Data Integration

This chapter provides an illustrated example of how you can implement the CDISC SDTM using SAS Clinical Data Integration 2.6. This chapter uses the metadata that was defined and loaded into the SAS Clinical Standards Toolkit as described in Chapter 4 as a prerequisite. First, you will see how to load that study metadata into SAS Clinical Data Integration. Then you will see how to set up your particular SDTM conversion study based on those standards and how to implement the SDTM conversions. Finally, you will be shown how to create define.xml using SAS Clinical Data Integration.

SAS Clinical Data Integration Introduction

For the clinical trials industry, the development of SAS Clinical Data Integration is a big step in the right direction. Instead of the historical technique of writing code by hand in Base SAS, SAS Clinical Data Integration provides a graphical interface to help you implement standard transformations based on the CDISC data standards. The graphical interface reduces the burden of generating code manually. The fact that SAS Clinical Data Integration is driven by the CDISC data standards helps ensure that the end product is compliant with the data standard.

The examples in this book use SAS Clinical Data Integration 2.6, which is built on top of SAS Data Integration Studio 4.9. SAS Data Integration Studio is a graphical interface-based ETL (extract, transform, load) tool similar to other ETL tools. The SAS Clinical Data Integration software is an add-on product to SAS Data Integration Studio to support those who work in clinical trials research and especially those who want to use the CDISC models. SAS has invested in helping the clinical trials research industry by developing SAS Clinical Data Integration as well as the complimentary SAS Clinical Standards Toolkit software shown in Chapter 4.

The use of SAS Clinical Data Integration itself could fill up a book and a course by itself, so this book cannot cover every aspect of its use. We will try to show as much detail as we can here, but some information has been left out to keep this book to a manageable size. Some setup topics have been left out, including but not limited to user management and authority settings, which are defined in SAS Management Console.

SAS Clinical Data Integration Metadata

This section shows how CDISC SDTM metadata is defined within SAS Clinical Data Integration. In Chapter 4, we saw how to use the SAS Clinical Standards Toolkit as a stand-alone product. However, the SAS Clinical Data Integration application is designed to work closely with the SAS Clinical Standards Toolkit. So we will now take our metadata definitions from Chapter 4 and use them here in Chapter 5 to feed the SAS Clinical Data Integration tool.

Classifications of SAS Clinical Data Integration Metadata

CDISC-Published Metadata

SAS Clinical Data Integration comes preloaded with the published CDISC standards as they exist when you install the software. That consists of the current standard domains for the published CDISC SDTM, CDISC-controlled terminology, and standard data classes for CDISC ADaM, CDISC SEND, the Operational Data Model, and define.xml. There are numerous metadata attributes, such as variable lengths in the CDISC SDTM, which are not defined by CDISC. Therefore, SAS makes a best guess as to what you might want for those. If you can use those data standards as SAS has defined them, then you are all set and you can start using SAS Clinical Data Integration very quickly.

Company-Specific Metadata

It is more likely that you have company-specific metadata based on the generic CDISC metadata that you will need to use in your data transformations to the SDTM. You might have a company-wide standard for SDTM implementations. You are also likely to have company-specific controlled terminology that you want to use. You might have drug compound standards or even study-specific standards that you want to use. If you work for a contract research organization, it is likely that you have numerous SDTM data standards to match the requirements of each sponsor organization. The good news is that SAS Clinical Data Integration enables you to have multiple data standards and to use them where you need them. In this book, we use the XYZ-SDTM-3.2 standard that we defined in Chapter 4 as our drug compound-specific SDTM standard, along with the default controlled terminology package loaded with SAS Clinical Data Integration. You can manually enter study metadata into SAS Clinical Data Integration. But it is easier to load the metadata into that tool via the SAS Clinical Standards Toolkit. Also, a newer feature of SAS Clinical Data Integration enables you to import standard metadata from a properly formed define.xml file. This can be especially helpful if you want to clone study metadata from another CDISC study into SAS Clinical Data Integration.

Setup of SAS Clinical Data Integration Metadata

You first need to import the XYZ-SDTM-3.2 standard that we defined in Chapter 4 as our drug compound-specific SDTM standard. To do this, go to the **Clinical Administration** tab and right-click the **Data Standards** folder to import the new standard.

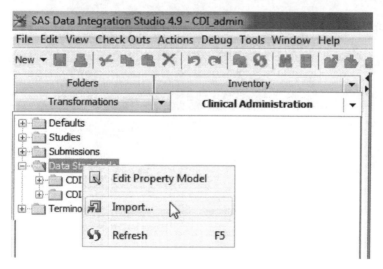

In the **Select Data Standard** window, select the SDTM standard type to import the custom standard that was registered in SAS Clinical Standards Toolkit. Here, you also see the CDISC-SEND and ADaM standards that came with SAS Clinical Data Integration.

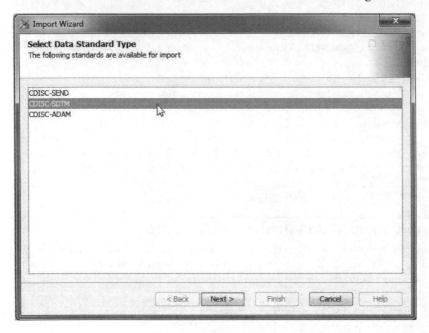

On the next screen, you see the new custom XYZ-SDTM-3.2 standard that you created in Chapter 4 as well as the other SDTM versions that the SAS Clinical Standards Toolkit has registered.

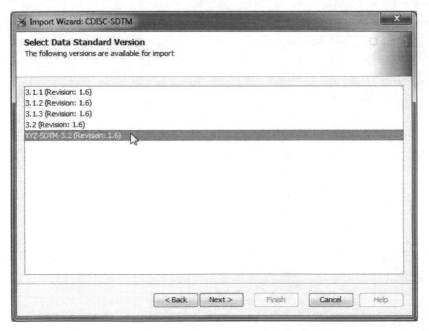

On the next screen, you define the general properties for the new custom XYZ-SDTM-3.2 standard. Most of that is populated with the data found in the SAS Clinical Standards Toolkit data. A few fields were modified, and the results look like this:

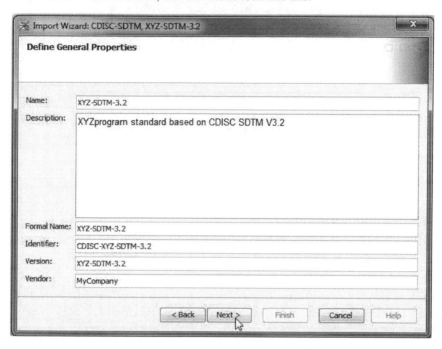

The screens that follow next—Verify Domain Properties, Verify Domain Column Properties, Validation Library, Verify Column Groups, Verify Columns in Column Group, Verify Domain Metadata, and Verify Domain Column Metadata—are left with the default values because they are populated by the SAS Clinical Standards Toolkit metadata for the XYZ-SDTM-3.2 standard from Chapter 4. If that work was not done as it was in Chapter 4, then SAS Clinical Data Integration would enable you to define that metadata on these screens. However, we suggest that you define your standards within the SAS Clinical Standards Toolkit in advance because there is a lot of metadata to edit, and editing those data on the fly on these few screens could be challenging. Once all of the metadata has been approved for use, register the XYZ-SDTM-3.2 standard into SAS Clinical Data Integration like this:

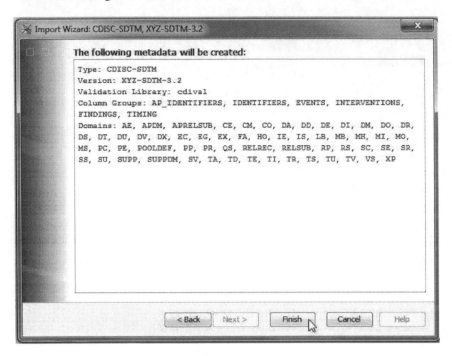

After you click **Finish**, the XYZ-SDTM-3.2 standard is registered into SAS Clinical Data Integration and can be used in SDTM data transformations. Notice how in the above screen capture the customized XP and SUPPDM domains have been loaded as well.

To use this standard, you must make it active. Go to the **Clinical Administration** tab, right-click the **XYZ-SDTM-3.2** drug-program-specific standard, and then select **Properties**. Under the **Properties** tab, set the **Active** status to **true**, like this:

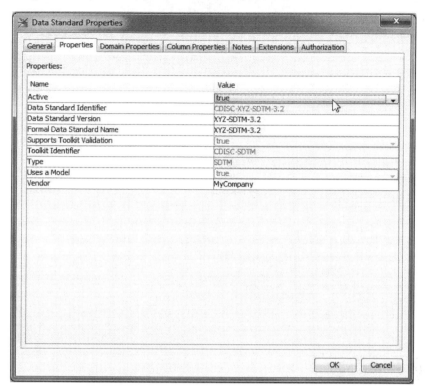

SAS Clinical Data Integration Study Setup

Before you start using SAS Clinical Data Integration for SDTM domain creation, a few study-specific setup steps are required. These are the items that you will want to do before beginning your data transformations to the SDTM:

1. Set up a clinical study object with standard subfolders to arrange your work.
2. Define SAS librefs to point at your source and target datasets.
3. Register your source datasets in SAS Clinical Data Integration.
4. Define your target datasets in SAS Clinical Data Integration.
5. Set general SAS Clinical Data Integration options.

Define the Clinical Study and Subfolders

To define a new study that you will call protocol XYZ123, you click the **New** tab and then select **Study**. The next screen is a General Information screen where you can enter a short name and a long name for the study. The next screen is the **Data Standards Selection** screen, which enables you to select the standard that is registered with SAS Clinical Data Integration that you want to use. Select the XYZ standard like this:

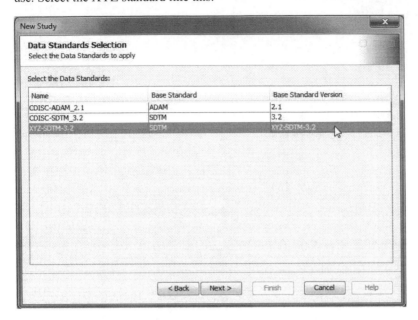

The following screens enable you to enter protocol information into the **Properties**, assign any SAS librefs in **Library Selection**, and assign a set of controlled terminology in the **Controlled Terminology** screen. For this study, just select the default SAS Clinical Data Integration Studio SDTM-controlled terminology package defined by CDISC and click **Finish**.

Now our new study XYZ123 is set up, and it looks like this:

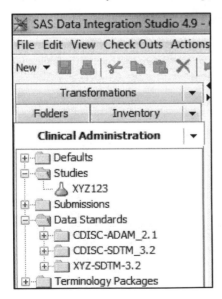

The next step is to define some objects within the study XYZ123. We want to define a folder for our SDTM work that we will call **SDTM**, a folder for our source data called **Source**, a folder for our ADaM work called **ADaM**, and three SAS librefs that point to those three locations. When we are done, our study folder now looks like this:

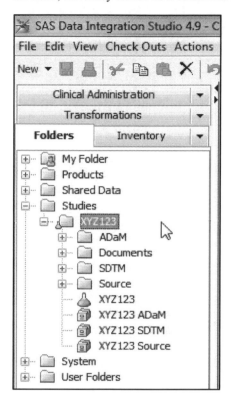

You create the study subfolders by right-clicking the study **XYZ123** study folder with the flask and select **New ▶ Folder**. The SAS librefs are defined by right-clicking the **XYZ123** study folder; selecting **New ▶ library**; and then following the prompts for the SAS libref name and physical location. The XYZ123 SDTM libref points to the **SDTM** folder, and the XYZ123 Source libref points to the **Source Data** folder. The **XYZ123** icon in the subfolder that is a flask represents metadata that is stored in the study folder that was created during the study creation process. You will not use this object directly.

Note that you can define a standard setup for how you define a new study within the **Clinical Data Administration** tab. Under the **Defaults** folder, you can define a type of template where you can create a study or submission template structure from which to copy and create new studies or submissions. For this book, we just defined a study object. However, you could create other object templates such as one for an integrated summary of efficacy, safety, or an NDA object. In this way, you can have a consistent definition for how you handle those other activities within SAS Clinical Data Integration as well.

Register Source Datasets and Define Target SDTM Datasets

The next step is to define your source data and target data within SAS Clinical Data Integration.

Register Source Datasets

The source data are registered when you right-click the **Source Data** folder and select **Register Data**. A Register Tables window appears, and you can select your data source like this:

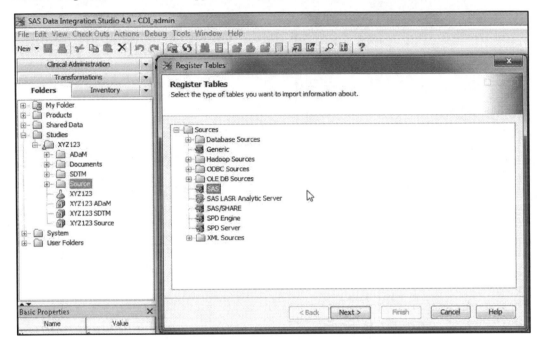

You then select the associated SAS library:

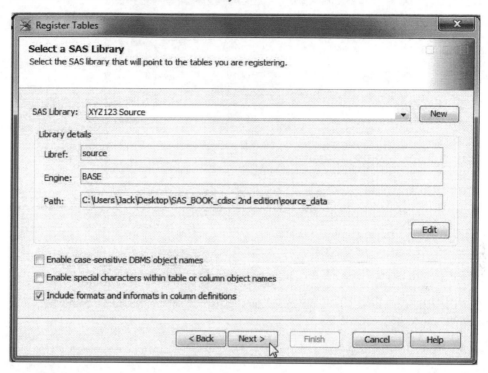

Then you select the source dataset that will be used in creating the SDTM domains. Note here that we have selected a dataset called FORMATDATA as well. This dataset is a SAS dataset copy of the CODELISTS.xlsx file from Chapter 2. We will use that codelist data in our lookup transforms later in this chapter.

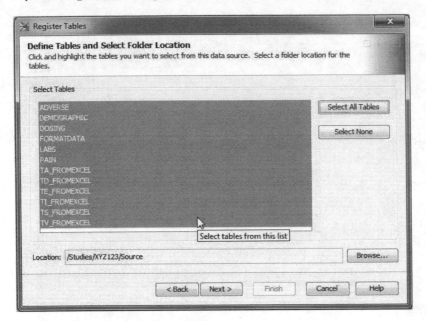

Click **Next** and then **Finish** to register these datasets. Now you should see the following within your study under the **Source Data** folder:

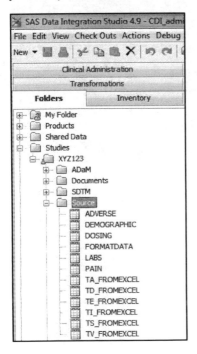

Define Target SDTM Datasets

Now that you have your source data visible in SAS Clinical Data Integration, you need to define your target SDTM datasets. Right-click the **SDTM** folder, select **New**, and then select **Standard Domain(s)**, like this:

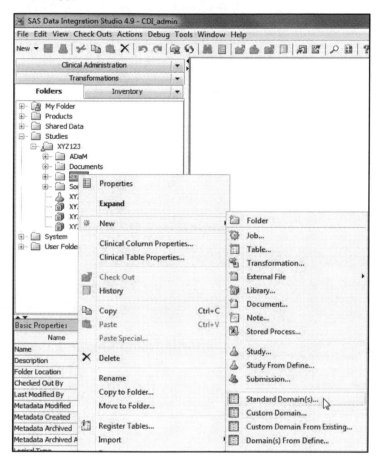

On the next screen, you are asked to select the domain location, which defaults to the folder selected. Then you have to assign the data standard to the domain like this:

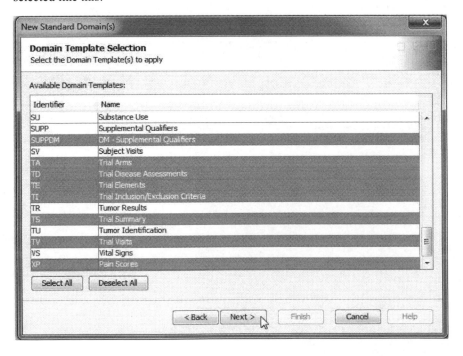

Then you select the SDTM domains that you want to define. Because the XYZ-SDTM-3.2 standard contains more domains than we need for this example, the individual sample domains are selected like this:

The target libref is then assigned like this on the Library Selection screen:

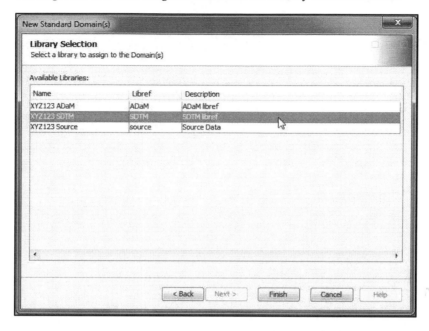

Finally, the target SDTM datasets are defined by the metadata and look like this in SAS Clinical Data Integration:

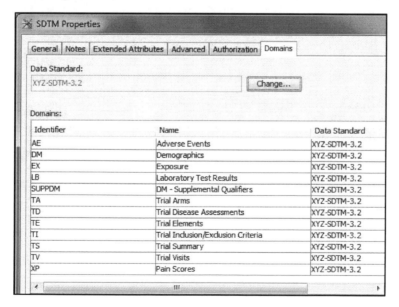

Setting SAS Clinical Data Integration Defaults

SAS Clinical Data Integration has some default operating behavior that can be specified. Much of this behavior can be customized to user tastes in terms of how things are displayed on the screen and tool default behavior. However, there is one default setting that you should absolutely change. Go to the toolbar, select **Tools**, and then select **Options**. You should see a screen like this:

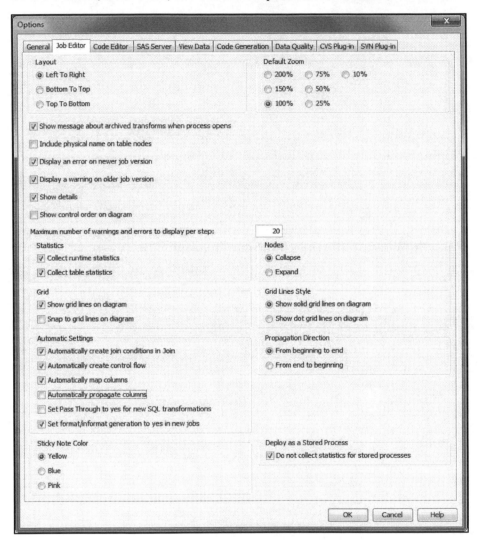

 If you click the **Job Editor** tab on the screen above, please be sure to turn off **Automatically propagate columns** under **Automatic Settings**. Automatic propagation **CAUTION** tells SAS Clinical Data Integration to automatically include variables upstream or downstream from the step that you are working on. This can have very bad unintended and unexpected consequences to the data transformations in your SDTM conversion work. Make sure you turn off auto propagation.

Creating SDTM Domains

Up to this point, you have performed the setup tasks for SAS Clinical Data Integration. You have set default options, created a study and associated standard, registered the source data, and defined the target SDTM domain data tables. It is now time to populate those SDTM data tables with data. This section shows you how to build the CDISC SDTM domains DM, SUPPDM, AE, XP, EX, LB, and some trial design datasets. In the course of those transformations, the most common SAS Clinical Data Integration transformations used are explored. The most commonly used transformations are the Extract, Table Loader, Transpose, SQL Join, Lookup, and Subject Sequence Generator. At the end of this section, we make a point about using customized SAS code, as well as discuss the concept of templating your jobs.

Creating the Special-Purpose DM and SUPPDM Domain

The first step in building the DM and SUPPDM SDTM datasets is to create a SAS Clinical Data Integration job. Right-click the **jobs** folder, and then select **New** and **Job**. A blank job diagram appears. The job diagram is a graphical user interface to which you can drag source and target data tables as well as action tasks that SAS calls transformations. Here is the completed job called "Make DM" that we will dissect:

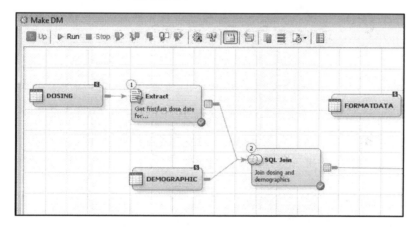

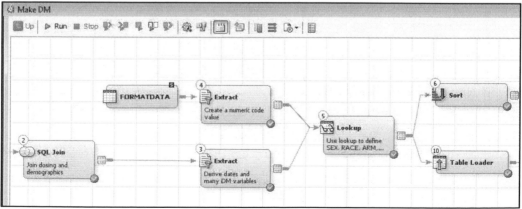

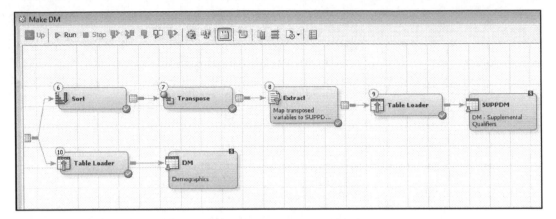

You can follow the job flow from the upper left to the lower right. In Step 1, an Extract transformation is used to take the raw dosing dataset and find the first date of dosing for the RFSTDTC and other dosing variables in DM. Step 2 joins that data with the source demographics dataset. Steps 3 to 5 use a Lookup transformation to define SEX, RACE, ARM, and ARMCD. Steps 6 to 8 use the Sort, Transpose, and Extract transformations to format the Other Race and Randomization Date data for SUPPDM use. Steps 9 and 10 use the Table Loader transformation to put the data content into the target SUPPDM and DM domain tables, respectively. We will look at some of these transformations in more depth now.

Extract Transformation

The Extract transformation is a common transformation that you can use to subset records and to keep, drop, and define new columns in the data table. Any transformation can be dragged from the **Transformations** tab in SAS Clinical Data Integration to the job diagram pane. If you drag a transformation and right-click it to select properties, the **Mappings** tab is the one where you do the majority of your data mapping work. For Step 1 in the above job, that **Mappings** tab looks like this:

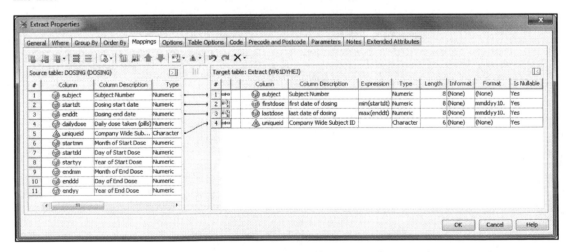

You see that we keep the SUBJECT and UNIQUEID variables as is. But look at the Expression column in the target table for FIRSTDOSE and LASTDOSE. Those variables are defined as the minimum of the dosing start date (FIRSTDOSE) and the maximum of the dosing end date

(LASTDOSE) for STARTDT and ENDDT respectively. The Expression text allows for any valid SAS SQL expression. You need to be fluent in SQL to use the Expression definitions effectively. For example, instead of a DATA step IF-THEN-ELSE statement for decision rule programming, you would use a CASE statement in SQL in the Expression text. In the above example, the expression derivation is small. If you needed a complex derivation, you could click in the Expression cell and the drop-down menu for an Advanced definition, which gives you a larger space to enter SQL expressions, or SAS will assist you with a list of SQL functions. Here is the Extract transformation from Step 3 that has a bit more mapping and definition in the Expression fields:

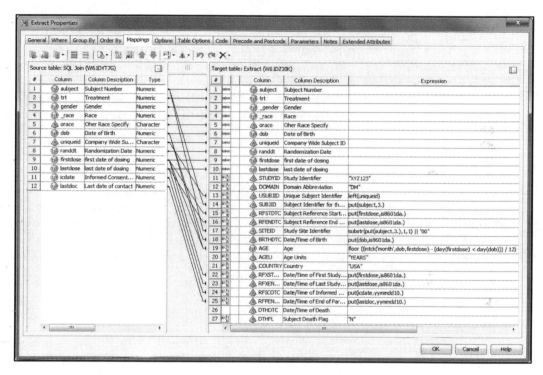

Note that the output dataset from this Extract Step 3 is a SAS work dataset called W61DZ20K. SAS Clinical Data Integration has a mechanism for naming datasets that results in system-derived dataset names that are not the easiest to remember. Fortunately, because you are not manually writing SAS code here, you generally do not have to know the names of the SAS work datasets. You can drag and make connections between data with your mouse.

SQL Join Transformation

The SQL Join transformation is used to join or merge datasets in SAS Clinical Data Integration. You see this transformation used in Step 2 of the job diagram above. After the SQL Join transformation has been dragged to the job diagram, you connect the datasets to be joined together. (In this case, these are the DEMOGRAPHIC dataset and the SAS work dataset from

Step 1 called W61DYHEJ.) Then the SQL join can be defined. If you double-click the SQL Join transformation, you see a window like this:

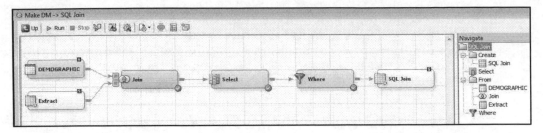

The Join box enables you to define what type of join (inner, outer, left, right, and so on) and what the join variables are. If you double-click the **Select** box, you essentially get the same window as the Mappings window from the Extract transform where you can choose which columns end up in your output dataset. Finally, you can use the **Where** box for a WHERE clause subsetting definition. Because the SQL Join transformation is your primary way to bring datasets together in SAS Clinical Data Integration, you want to make sure you are fluent in SAS SQL to leverage the strengths of this transformation.

Lookup Transformation

The Lookup transformation in SAS Clinical Data Integration enables you to create a variable based on a source variable that is then compared and mapped to a codelist table. You could almost think of this as the equivalent of applying a SAS format to a source variable to create a new variable. We will explore this part of the job diagram here:

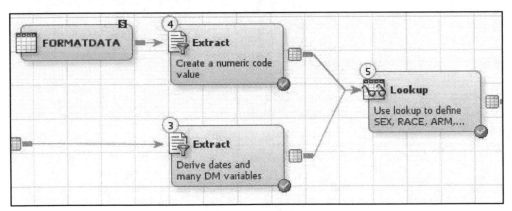

The FORMATDATA dataset is the SAS dataset equivalent of CODELISTS.xls from Chapter 3 because we are using the same lookup data. The extract step in Step 4 was necessary to create a numeric code value to compare with the numeric source data from Step 3. As a refresher, here is what that dataset looks like at the end of Step 4:

#	fmtname	start	end	label	type	numeric_code
1	ACN_adverse_aeaction	3	3	DOSE INCREASED ...	N	3
2	ACN_adverse_aeaction	4	4	DOSE NOT CHANGED ...	N	4
3	ACN_adverse_aeaction	2	2	DOSE REDUCED ...	N	2
4	ACN_adverse_aeaction	1	1	DRUG INTERRUPTED ...	N	1
5	ACN_adverse_aeaction	5	5	UNKNOWN ...	N	5
6	AEREL_adverse_aerel	1	1	NOT RELATED ...	N	1
7	AEREL_adverse_aerel	2	2	POSSIBLY RELATED ...	N	2
8	AEREL_adverse_aerel	3	3	PROBABLY RELATED ...	N	3
9	AESEV_adverse_aesev	1	1	MILD ...	N	1
10	AESEV_adverse_aesev	2	2	MODERATE ...	N	2
11	AESEV_adverse_aesev	3	3	SEVERE ...	N	3
12	ARM_demographic_trt	1	1	Analgezia HCL 30 mg ...	N	1
13	ARM_demographic_trt	0	0	Placebo ...	N	0
14	ARMCD_demographic_trt	1	1	ALG123 ...	N	1
15	ARMCD_demographic_trt	0	0	PLACEBO ...	N	0

View Data: Extract (56 rows)

The Lookup transformation is designed so that you can perform multiple codelist lookups in a single transformation. In Step 5 above, you can right-click the Lookup transformation to add new input ports. This was done here for the three additional lookups. We will now explore the DM domain RACE derivation, which involves mapping the numeric _RACE source variable with the FORMATDATA dataset. You have three items to define here. A WHERE clause, a source to look up, and a lookup to target mapping. If we right-click the Lookup transformation and select **Properties**, there is a **Lookups** tab where the lookup mappings are defined. If we select the first lookup for RACE, we can click **Lookup Properties** and the **Where** tab and we see this:

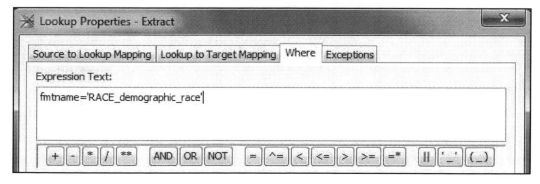

This subsets your table lookup for where the SAS format name was RACE_demographic_race, which is the one to specify for the DM RACE variable mapping. Then, if you click **Source to Lookup Mapping**, you see:

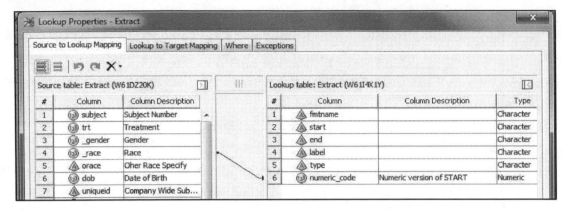

In the lookup table, _RACE is mapped to NUMERIC_CODE. Finally, if you click **Lookup to Target Mapping**, you see:

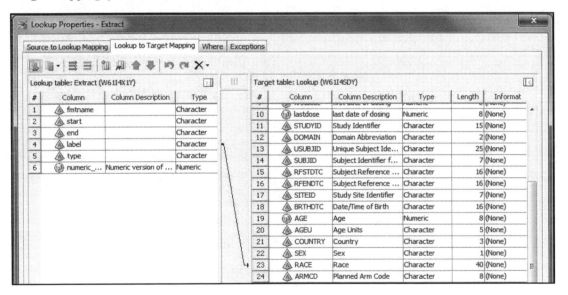

Here you can see that the LABEL variable that holds the decoded RACE value is mapped to the eventual RACE variable in the DM domain. This lookup process is repeated for SEX, ARM, and ARMCD, as well as in the Lookup transformation Step 5.

Transpose Transformation

Because the SDTM contains datasets that often consist of name/value pairs, you will spend a lot of time flipping or transposing your data structures. In SAS Clinical Data Integration, the tool for flipping your data on end is the Transpose transformation. We will look at that here in the process of creating the SUPPDM qualifiers for other races and the randomization date. In our source datasets, these data are columns. But in SUPPDM, they exist as rows. If you right-click the

Transpose transformation in Step 7 of the job diagram and then select the **Options** tab, you see a screen like this:

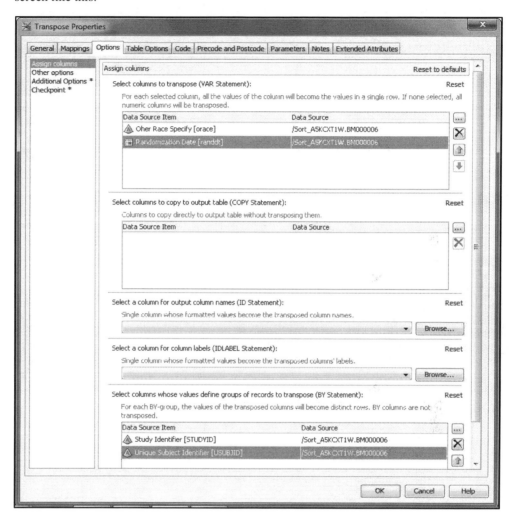

What you see here are the typical options that you have when you write PROC TRANSPOSE programming code. We have a simple case here, so ORACE and RANDDT are selected as the columns to transpose, and STUDYID and USUBJID are selected as the variables to transpose by. Then, if you click the **Mappings** tab, you see this:

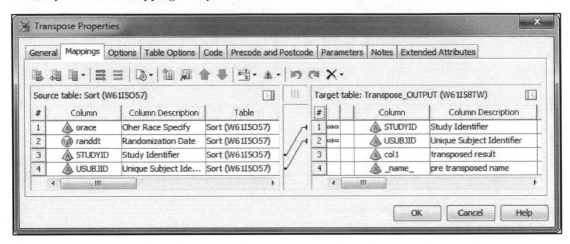

You can see that STUDYID and USUBJID are mapped straight over to the target table. However, notice the new COL1 and _NAME_ variables. If you are familiar with PROC TRANSPOSE, you might recognize these as common output variables from the TRANSPOSE procedure. Unfortunately, these variables are not automatically generated by this transformation. You need to tell SAS what these variables are, and you have two choices for how to do it. One option is to insert a User Written Code transformation and do a PROC CONTENTS on the resulting W61158TW dataset to see what PROC TRANSPOSE generates. Then, you go back and update the target mapping with the appropriate metadata for COL1 and _NAME_. The other option is to have the Transpose transformation automatically update the output dataset metadata for you. You can do this on the Transpose transformations Options tab by selecting **Additional Options** and changing the **No** value shown in the following screen to **Yes**:

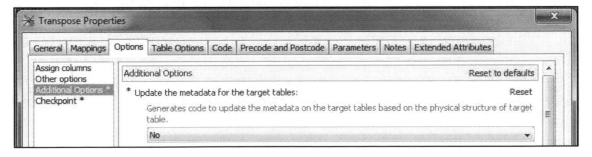

Hopefully, future iterations of the Transpose transform will make this task a bit easier, since data transposition is common in SDTM data creation. Now that ORACE and RANDDT have been transposed from columns to row values, the only task left is to map those values to the appropriate columns in SUPPDM. That Extract transformation, which is found in Step 8, looks like this:

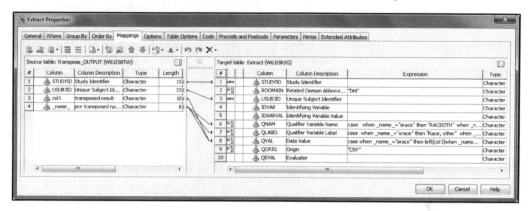

Notice how variables COL1 and _NAME_ are used to populate QNAM, QLABEL, and QVAL with more complex CASE statements in the **Expression** field. Here is what that advanced Expression window looks like for the QVAL definition:

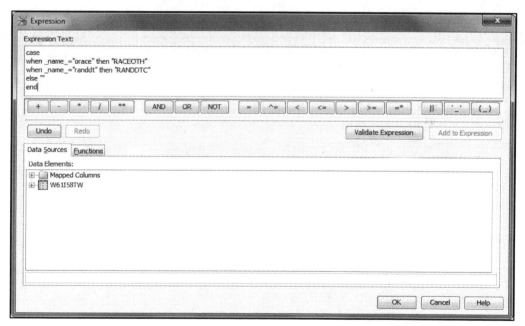

At this point, it is just a matter of dragging the Table Loader transformations to the job diagram and loading the data from Steps 9 and 10 into the target SUPPDM and DM datasets respectively.

Creating the AE (Adverse Events) Events Domain

The next SDTM domain to explore is the Adverse Events (AE) events model domain. The good news with most events-based data is that, in general, it is probably structured in your source data in a one-record-per-event data structure, similar to the structure that is needed in the SDTM events structure. The AE job diagram for study XYZ123 in SAS Clinical Data Integration looks like this:

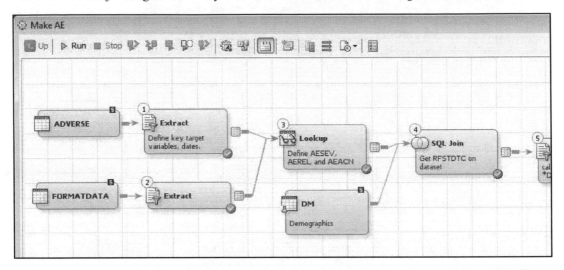

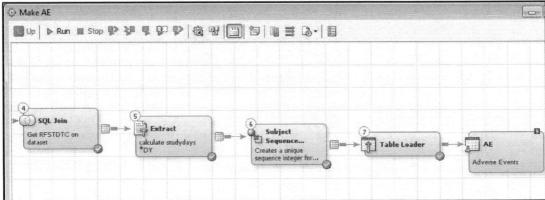

In Step 1 of the diagram, the source dataset ADVERSE has a number of the key SDTM AE variables defined in the Extract transformation. The Lookup transformation is used in Step 3 much as it was in the DM job so that the AE SDTM variables AESEV, AEREL, and AEACN could be created based on a lookup table. Step 4 joins the adverse event data with the DM domain to get the RFSTDTC variable, which is needed to define SDTM study-day variables AESTDY and AEENDY in Step 5. The Advanced expression for AESTDY in Step 5 looks like this:

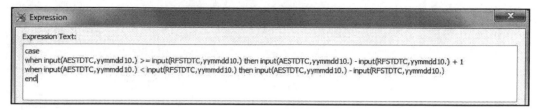

AEENDY is defined in the same way, with AESTDTC being replaced by AEENDTC above. Keep in mind that this expression works for dates only. If you had time components, then those would need to be added to the expression as well.

Subject Sequence Generator Transformation

SAS supplies you with a few SDTM-friendly custom transformations with SAS Clinical Data Integration. The Subject Sequence Generator transformation is a pre-built transformation supplied by SAS that defines SDTM --SEQ variables for you. You can drag this transformation from this location under the **Transformations** tab:

Drag that transformation to the make AE job as Step 6. If you double-click that transformation and select the **Options** tab, you see this:

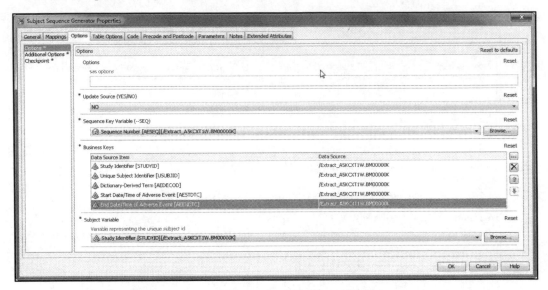

You need to select the --SEQ variable that you are trying to define, which should exist as a blank variable in the incoming dataset. In this example, that is AESEQ. Then, you need to define the **Business Keys**, which is the sort sequence that you want to use to define the record order for AESEQ. Finally, you need to identify the variable that you want to initialize AESEQ with; in this case that is STUDYID. The Subject Sequence Generator transformation populates AESEQ on the resulting dataset.

This job ends with Step 7, which uses the Table Loader transformation to push the working dataset into the final SDTM AE dataset.

Creating the XP Pain Scale Customized Findings Domain

The custom XP pain scale data findings domain for study XYZ123 is probably the longest job in this study. This domain introduces us to a new data transformation issue that we will explore. The XP domain is a customized findings domain that was defined in the SAS Clinical Standards Toolkit and brought into SAS Clinical Data Integration to be populated for study XYZ123. Here is what that job diagram looks like from left to right and top to bottom:

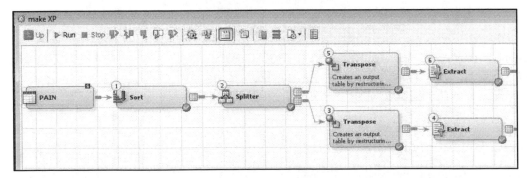

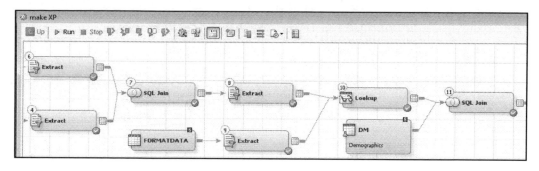

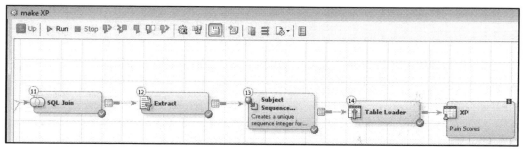

The problem with the PAIN dataset is that it is somewhat flat in nature in that the three pain measurements and dates all exist on the same row. The PAIN source data look like this:

#	subject	randomizedt	month3dt	month6dt	painbase	pain3mo	pain6mo	uniqueid
1	101	04/02/2010	07/03/2010	10/10/2010	3	2	1	UNI101
2	102	02/13/2010	05/10/2010	08/11/2010	3	3	1	UNI102
3	103	05/15/2010	08/15/2010	11/15/2010	3	3	0	UNI103
4	104	01/02/2010	04/03/2010	07/04/2010	2	0	0	UNI104
5	105	04/20/2010	07/20/2010	10/19/2010	3	3	1	UNI105
6	106	04/01/2010	07/05/2010	10/10/2010	3	0	0	UNI106
7	201	06/10/2010	09/09/2010	12/12/2010	3	2	0	UNI201

View Data: PAIN (60 rows)

To structure this data like we need it for the XP domain, we need to split the data, transpose it, and then join it back together on the other side. This processing of the data is handled from Step 2 to Step 7 in the job diagram. Step 2 creates one work dataset that holds the pain measurement date variables RANDOMIZEDT, MONTH3DT, and MONTH6DT, and another work dataset that contains the pain measure variables PAINBASE, PAIN3MO, PAN6MO. Steps 3 and 5 flip or transpose that data so that those columns become rows. Steps 4 and 6 create an INDEX variable that goes from 1-3, based on the PROC TRANSPOSE variable called _NAME_, so that Step 7 can join the data back together.

The rest of this job is similar to what we have seen in previous jobs. Step 8 is a simple Extract transform that maps INDEX to VISITNUM and creates columns for XPSEQ, XPTESTCD, XPTEST, VISIT, and XPDY that will be defined downstream. Step 10 is used to create VISIT with the Lookup transform. Steps 11 and 12 are there to join the pain data with the SDTM DM file so that the study-day XPDY can be created. Step 13 creates XPSEQ with the Subject Sequence Generator transform, and Step 14 loads the pain data into the XP custom domain.

Creating the EX Exposure Interventions Domain

The next SDTM domain to explore is the EX exposure model domain. Here is what that job diagram looks like:

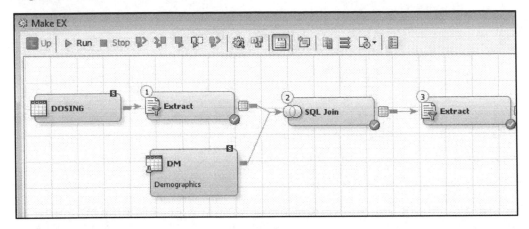

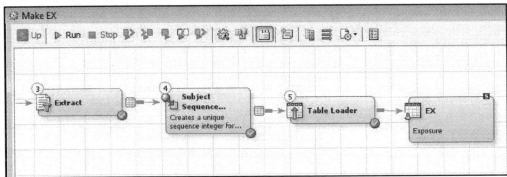

This is a fairly simple domain to generate in SAS Clinical Data Integration. Step 1 derives USUBJID, EXDOSE, EXDOSTOT, EXSTDTC, and EXENDTC. In Step 2, the temporary dataset from Step 1 is joined with the DM domain to get RFSTDTC so that EXSTDY and EXENDY can be derived in Step 3. Step 4 is the Subject Sequence Generator transform call, which creates EXSEQ. Finally, the dataset is loaded into the EX target domain in Step 5.

Creating the LB Laboratory Findings Domain

The last SDTM domain to look at that requires any real data manipulation is the LB domain. Here you will see transformations similar to ones that you have seen before. The job diagram for LB in SAS Clinical Data Integration looks like this:

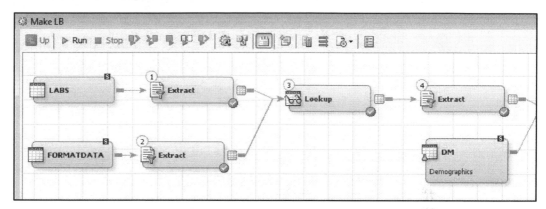

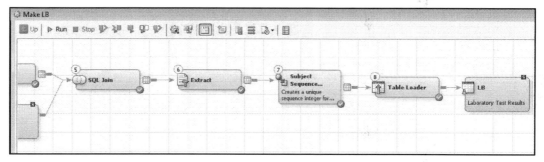

As you have seen before, in Step 1 are the USUBJID target variable and other LB target variables. Steps 2 and 3 use the Lookup transform to define VISIT, LBCAT, LBTEST, and LBTESTCD, based on our controlled terminology. Step 4 defines baseline lab data in LBBLFL and reference range flagging in LBNRIND via fairly simple CASE statements. Steps 5 and 6 bring in the DM RFSTDTC variable so that LBDY can be created. Finally, LBSEQ is created in Step 7, and Step 8 loads the dataset into the LB domain.

Creating the Trial Design Model Domains

The majority of the Trial Design Model domains in the SDTM are pure study metadata and include no actual patient data. In this book, we did not derive the Subject Elements (SE) and Subject Visits (SV) domains, which are based on the subject data. For this book, we generated Trial Arms (TA), Trial Disease Assessments (TD), Trial Elements (TE), Trial Inclusion (TI), Trial Summary (TS), and Trial Visits (TV), which are trial metadata domains. In Chapter 3, we entered that information into Microsoft Excel spreadsheets and imported it into SAS with PROC IMPORT. For SAS Clinical Data Integration, we could do the same thing, assuming that the machine that has SAS Clinical Data Integration on it has SAS/ACCESS Interface to PC Files

installed. For this book, we simply converted those spreadsheets into SAS datasets called
**_FROMEXCEL and then registered those datasets as source data. The job diagram to create
TA, TE, TI, TS, and TV looks like this:

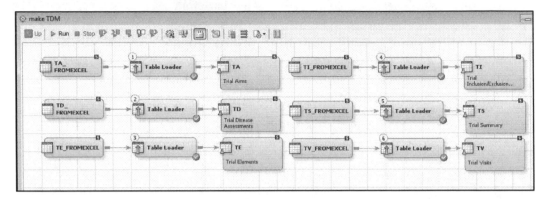

This job is then just a matter of loading the six source datasets into the six target SDTM domains.

Using Customized Code in SDTM Production

SAS Clinical Data Integration has a User Written Code transform that enables you to create your
own handwritten code if necessary. It is advisable to use this transform sparingly if at all because
the manipulations that you do in it will generally be outside of the metadata control of the rest of
your SAS Clinical Data Integration job. For example, if you write custom DATA step code and
you define lengths and attributes for variables there, then SAS Clinical Data Integration will not
be aware of those manually entered metadata elements. We suggest that you use the User Written
Code transform for data and job checking and monitoring code, but otherwise you should stay
with using the provided and metadata-controlled transforms.

There is a type of user-written code in SAS Clinical Data Integration that you should take full
advantage of. SAS Clinical Data Integration gives you the ability to write your own user-written
transformations. These transforms can be written so that they are metadata- and parameter-driven,
just like the other SAS Clinical Data Integration transforms. An example of a user-written
transform is the Subject Sequence Generator transform that is supplied by SAS. You can explore

that to see how one of these transforms is created. If we drill down under the **Clinical** transformations and examine the **Properties** and **Source Code** of that transform, you see this:

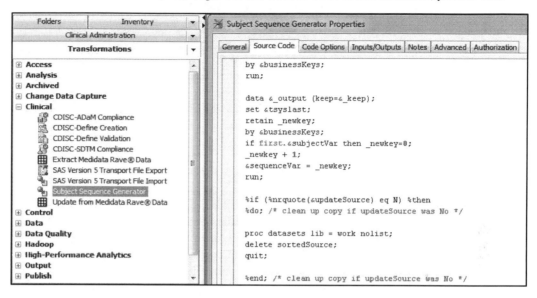

You can see the DATA step code that defines the **SEQ SDTM variable in the &sequencevar macro variable. The majority of the code that SAS Clinical Data Integration generates is SQL-based code, but creating the **SEQ variable is easy to do using DATA step mechanics. The code that you see here is very much like what we saw in Chapter 3 in the solution that used only Base SAS. You can think of these SAS Clinical Data Integration user-written transformations as generic SAS macros that are metadata- and parameter-driven. If you wanted to create a new transformation, you can right-click in the folders area and select **New Transformation** to begin the process of defining a new user-written SAS Clinical Data Integration transform.

If you find yourself performing repetitive tasks in SAS Clinical Data Integration, then that is a good time to think about writing your own user-written transform. The Subject Sequence Generator is an obvious candidate that SAS has written for you. In the domains generated in this book, adding a transform that can easily join a source dataset with the SDTM DM file for study-day (**DY) variable generation would be a good choice. Another good automatic transform is one that automatically generates common variables such as USUBJID, STUDYID, or even **DTC dates. The nice thing is that if you have a repetitive task, then SAS Clinical Data Integration gives you a way to create a data-driven generic SAS macro or user defined transform to perform that task.

Templating Your SDTM Conversion Jobs for Reuse

You might want to reuse some of your SDTM data conversion jobs within SAS Clinical Data Integration. Perhaps you have two similar studies, and you want to copy a DM job from one study to the other to save time. To copy a job easily from one study to another, you will want to construct it in a certain way. Let's call this construction templating your job. You might have noticed that many of the SDTM domain conversion jobs previously mentioned have a general approach that look like the following figure.

Figure 5.1: General SAS Clinical Data Integration Job Flow to Create an SDTM File

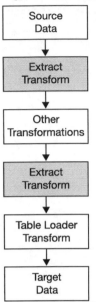

Note the Extract transformations shaded in the figure. These Extract transformations tend to be straight data mapping from the object on the top to the object on the bottom. These steps save the incoming and target variable lists so that any mappings or other processing are not lost when the sources and targets are changed. In addition, it makes any variable changes clear when applying new source or target tables. If you create your jobs in this fashion, then you can copy them from one place to another and simply change the source and target data. The Extract transformations will protect the work in between.

Using SAS Clinical Data Integration to Create Define.xml

When all of the SDTM domains have been generated in SAS Clinical Data Integration, you might want a define.xml file. Fortunately, there is an automated SAS Clinical Data Integration transformation called `CDISC-DEFINE CREATION` that performs this task. Here is the job diagram for creating the define file:

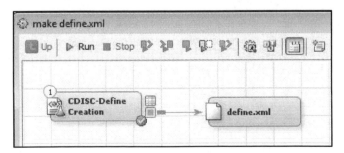

As simple as that looks, there are a few items that need definition before the define file can be created. If you right-click **CDISC-DEFINE CREATION**, you see that you can select the data standard, the study, and the specific domains that you want to be in your define file:

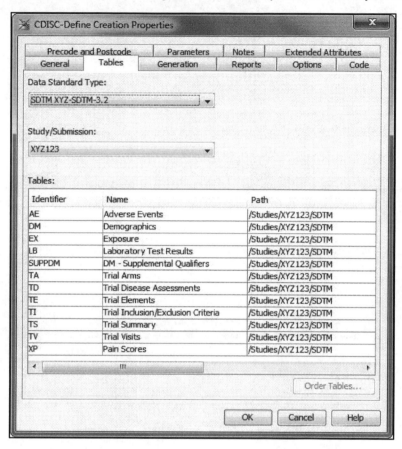

The **Generation** tab provides an option for using no style sheet, using the generic CDISC-supplied style sheet, or using your own custom style sheet. You can also select the version of define.xml that you would like. In this case, we chose define version 2. The only other piece to define is the actual define.xml file itself. That file can be defined by right-clicking the target SDTM folder and selecting **New Document**. Then, the output port on the CDISC-DEFINE

CREATION transform can be connected to the define.xml file. You can see that connection in the job diagram, as well as the folders view on the left:

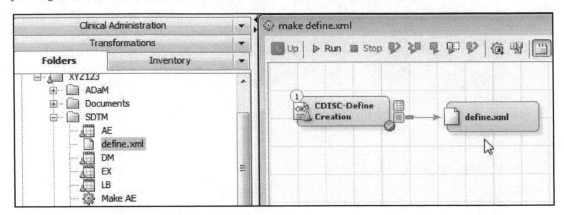

The result of this job is a define file and a style sheet called define2-0-0.xsl. A snapshot of the define file that is generated by SAS Clinical Data Integration looks like this:

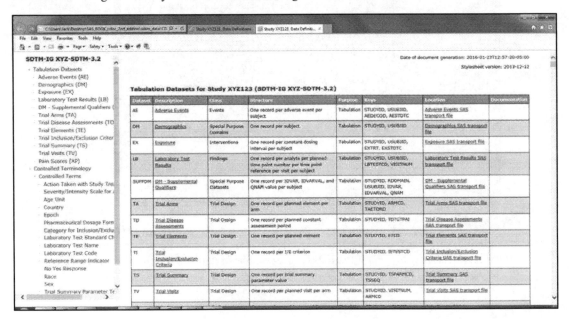

Chapter Summary

This chapter described how you can use SAS Clinical Data Integration to create CDISC SDTM files and how to generate a define.xml file. The SAS Clinical Data Integration examples in this chapter were built on top of the SDTM metadata defined in Chapter 4. The metadata (found in the SAS Clinical Standards Toolkit) from Chapter 4 was used to populate the target SDTM datasets. Then we used the graphical user interface of SAS Clinical Data Integration to primarily point and click our way to SDTM production. The final task was to use a transformation provided by SAS to build the define.xml file. This general process can also be followed in SAS Clinical Data Integration to create ADaM datasets and define files as well.

Chapter 6: ADaM Metadata and ADaM Define.xml

As discussed in Chapter 1, one of the fundamental principles of ADaM analysis datasets is that they be accompanied by metadata. The ADaM 2.1 model document contains a full section on ADaM metadata, including illustrations of concepts such as analysis dataset metadata, analysis variable metadata, analysis parameter value-level metadata, and analysis results metadata. If you are familiar with SDTM metadata, then many of these ADaM metadata concepts will be familiar. They serve the same purpose of helping to describe the data or, in the case of analysis results metadata, describing the relationship between the analysis data and the analysis results.

In this chapter, we take what we learned in Chapter 2 about SDTM metadata and apply it to ADaM metadata and the creation of an ADaM define.xml file using a Base SAS implementation. The CDISC Define-XML version 2.0 release package contains a sample ADaM implementation with two ADaM datasets and a dummy reviewer's guide.   Building off of that is an even more complete example packaged as a part of CDISC's Analysis Results Metadata specification. The style sheet that comes packaged there will be used for the implementation to be demonstrated in this chapter. Both packages are available from the CDISC website, www.cdisc.org.

More Information

Appendix C – ADaM Metadata

The %make_define2 SAS macro and the ADaM metadata spreadsheets used for this book can be found on the authors' pages at http://support.sas.com/publishing/authors/index.html.

See Also

Define-XML version 2.0 (http://www.cdisc.org/standards/foundational/define-xml.)

Analysis Results Metadata version 1.0 for Define-XML v2.0 (http://www.cdisc.org/standards/foundational/adam)

Analysis Data Model version 2.1 (http://www.cdisc.org/standards/foundational/adam)

Metadata Spreadsheets

As stated in Chapter 2, good up-front SDTM metadata is essential for driving the entire SDTM implementation process. The same holds true for an ADaM creation process. Fortunately, many of the spreadsheets created for SDTM metadata can be used for ADaM metadata as well. We do not revisit those here, but if you are interested in seeing the specific ADaM metadata for our implementation, you can find it in Appendix C. However, two of the spreadsheets, VARIABLE_METADATA and VALUELEVEL_METADATA, deserve additional details for an ADaM implementation and are covered in the following sections. One set of metadata, analysis results metadata, is completely unique to ADaM and is covered for the first time in "*Analysis Results Metadata.*"

We should mention the existence of one additional column called DOCUMENTATION that is used in ADaM dataset metadata (which is captured in the TOC_METADATA spreadsheet) but not in the SDTM dataset metadata. It is here that you can provide further details of how the dataset was derived or additional references to protocols or analysis plans.

Variable Metadata in ADaM

Variable metadata in SDTM and ADaM are very similar. There are, however, fields that exist in one but not in the other. The columns for ROLE and ROLECODELIST, for example, have specific purposes for SDTM data but not for ADaM. They can therefore be ignored when you are using the spreadsheet for ADaM data. For your own implementation, you might choose to remove them altogether.

The updates in Define-XML version 2.0 are in many ways very friendly to the needs of ADaM metadata. In addition, there are also changes to the new style sheets that contribute equally as well to efforts to standardize how ADaM metadata should be captured and represented. Although the style sheet is not a part of the standard, it is important to discuss the two together since what people actually see when they view the metadata is determined by the style sheet.

One difference between the presentation of SDTM and ADaM variable metadata is with the Origin element. In SDTM, Origin is presented as its own column in data definition tables (DDTs). In ADaM, it appears under one column of Source/Derivation/Comment. This is largely due to specifications in the ADaM version 2.1 model document. One possible reason for the use of "Source" in ADaM metadata rather than "Origin" may be due to the ADaM fundamental principle that (emphasis added) "analysis datasets and their metadata should provide traceability between the analysis data and its *source* data." For practical reasons, the Origin metadata element is used to capture the ADaM variable's source, but what is represented in the Origin field in ADaM metadata should be the immediate predecessor of a given variable rather than its "genesis" origin. This is an example of how the define 2.0 standard has become friendlier to ADaM metadata: Among the list of controlled terminology for the Origin element's type is "Predecessor."

In our ADaM metadata and our %make_define2 macro, any Origin value that does not start with other values from the controlled terminology list ("CRF", "Derived", "Assigned", "Protocol", and "eDT") is assumed to be a predecessor. Any predecessor value that does not contain a period assumes that only the source dataset is being provided, and the variable names between the current dataset and the source dataset are the same. The macro therefore uses the existing variable name to complete the predecessor value. For example, AGE in ADEF can have an origin of just "ADSL," and the macro will complete this to "ADSL.AGE" in the define file. This is intended to save typing and reduce errors. If, however, the predecessor has a different variable name, then the full origin (with a period between the source domain/dataset and the source variable) should be specified.

In addition to "Predecessors", one of the most common origin types in ADaM metadata may be "Derived". The new style sheet presents derived fields in the Source/Derivation/Comment DDT column as "Derived:", and appends computational algorithm metadata to come afterward. The colon provides good motivation for ensuring that your metadata are complete with such details, although admittedly, such "derivations" can sometimes be difficult to specify succinctly with pseudo-code or other methods. As we will see later in the validation chapters, such derivations are now expected and can generate validation errors if not provided.

If a particular variable requires parameter value-level metadata, then there is no need to specify this explicitly in the variable metadata. All that is required is that the dataset and variable name be provided appropriately in the value-level metadata spreadsheet. This will ensure that the information is properly linked between the two spreadsheets. In our simplified examples, both BDS-structured datasets (ADEF and ADTTE) have only one parameter. So while the parameter value-level metadata is not necessary, it will be applied in order to demonstrate the functionality. In order to avoid confusion, information such as a codelist and computational algorithm will not be provided in the variable metadata for variables such as AVAL in BDS-structured datasets (ADEF and ADTTE in our example). Rather, just a comment to "Refer to parameter value-level metadata" is provided for these variables.

Analysis Parameter Value-Level Metadata

An analysis dataset that follows the ADaM Basic Data Structure (BDS) can contain many analysis parameters. The analysis variable in that structure, typically AVAL or AVALC, can have different attributes depending on the parameter. For example, as shown in Chapter 2, values for different lab tests will likely require different formatting. Some tests might be displayed only as whole numbers, and others might need three decimal places in order to capture meaningful differences between values.

The analysis parameter value-level metadata can go a bit beyond just describing attributes of analysis variables. With the possibility to add numerous derived columns to a BDS dataset, many of which can vary in their derivation or meaning depending on the given parameter, the supporting metadata can get rather intricate. To demonstrate this, consider an analysis dataset, ADEF, that contains efficacy results derived from the SDTM QS domain. In our simplified example data, we have only one questionnaire test, the pain question. This translates to an analysis parameter in our BDS-structured ADEF dataset. The analysis variable, AVAL, captures the response to the pain question; BASE captures the baseline value; and CHG captures the change from baseline. The primary endpoint in the study is a responder definition where a subject is considered a responder if he or she has an improvement in the pain scale of two or more points at 6 months. In the ADEF dataset, we use the variable CRIT1FL to indicate whether a subject met this responder definition at the given visit. CRIT1 is a text string to describe the result captured by CRIT1FL.

In theory, each parameter could have a different primary set of criteria that should be flagged. For example, other pain scales might be less sensitive, and a 1-point improvement in the scale might be considered clinically noteworthy. The value of CRIT1 would then vary depending on the parameter, and the derivation of CRIT1FL would also differ from the derivation for the primary pain question. These differences would have to be documented at the parameter level. Even with one parameter, documenting these differences at the parameter level would be the proper way to do it. This documentation for our example is shown in Table 6.1.

Table 6.1: Selected Analysis Parameter Value-Level Metadata for ADEF

VARIABLE	VALUEVAR	VALUENAME	TYPE	LENGTH	COMPUTATION METHODOID	CODELISTNAME
AVAL	PARAMCD	XPPAIN	Integer	8		PAINSCORE
CRIT1	PARAMCD	XPPAIN	Text	51		ADEF.CRIT1
CRIT1FL	PARAMCD	XPPAIN	Text	1	RESPONDER	YN

In order to make the data entry easier and less error prone, only the dataset name and variable name need to be entered into the value-level metadata spreadsheet. From those, the %make_define2 macro will create a VALUELISTOID that is used to uniquely identify each row in the define file. This is done as a simple concatenation of the two fields, separated by a period.

Using the same metadata fields shown in Chapter 2 for the SDTM, we can describe details at the parameter level that could not be shown otherwise. In rows where VARIABLE=AVAL, details for each pain parameter can be provided. In rows where VARIABLE=CRIT1, the length for each value of CRIT1 and the codelist that contains each unique value can be shown. Finally, rows where VARIABLE=CRIT1FL can display the computational method that might be unique for the given parameter, as well as codelists that might be parameter-dependent. Providing analysis parameter value-level metadata for multiple variables is more robust than the standard approach for SDTM value-level metadata. In the standard approach, only details pertaining to --ORRES variables for a given value of --TEST or --TESTCD are provided. With ADaM analysis parameter value-level metadata, details pertaining to essentially any variable for a given value of PARAM or PARAMCD can be provided.

Not shown in Table 6.1 is the WHERECLAUSEOID field. For simple value-level metadata, where the where clause condition is simply based on specific values of PARAMCD (as indicated by the VALUENAME values), manually entering the where clause (and coming up with unique where clause OIDs) would be cumbersome. The %make_define2 macro was therefore designed to programmatically create these simple where clauses. The field exists in the spreadsheet in case a more complicated where clause is needed. Otherwise, if the field is left blank, the macro will construct the where clause for you. If you are interested in seeing examples of more involved where clauses, see the analysis results metadata in Appendix C.

In a later section, we will show how this information is represented in the define.xml file.

Analysis Results Metadata

A unique component to ADaM metadata compared to its SDTM counterpart is analysis results metadata. Analysis results metadata contain some information pertaining to the study results that you might find in the clinical study report (CSR). Most ADaM metadata provide the traceability that you need to understand the data's lineage from SDTM to ADaM. But analysis results metadata provide the traceability that you might need to understand how the ADaM data are used to produce some of the key results that appear in a CSR. Although many might think of it as information that would be provided after analyses have been completed, it can be used, in conjunction with the statistical analysis plan (SAP), to provide SAS programmers with some details and specifications needed to carry out analyses.

The metadata captured in the ANALYSIS_RESULTS worksheet is described in Table 6.2.

Table 6.2: Description of Metadata Captured in the ANALYSIS_RESULTS Worksheet

Excel Column	Description
DISPLAYID	This is a unique identifier for the display (that is, a table, figure, or listing) that contains the analysis result. It is often populated with values that correspond to the number of the output table in the CSR. If you want to provide links from the define file to the CSR table, the necessary details must be provided in the EXTERNAL_LINKS sheet, and the DISPLAYID here must match the LEAF_ID in that sheet.
DISPLAYNAME	A title for the display of the analysis results. Often a table or figure title.
RESULTNAME	A text description of the analysis result. One analysis result can appear in multiple displays, and one display can contain multiple analysis results.
REASON	The rationale for performing this analysis. It indicates when the analysis was planned. Extensible controlled terminology includes: `SPECIFIED IN PROTOCOL`, `SPECIFIED IN SAP`, `DATA DRIVEN`, and `REQUESTED BY REGULATORY AGENCY`.
PURPOSE	The purpose of the analysis within the body of evidence (for example, a section in the clinical study report). Extensible controlled terminology includes: `PRIMARY OUTCOME MEASURE`, `SECONDARY OUTCOME MEASURE`, `EXPLORATORY OUTCOME MEASURE`.
PARAMCD	The PARAMCD value to which the analysis applies, if applicable
ANALYSISVARIABLES	The analysis variables to be analyzed. Often AVAL, CHG, or both in BDS-structured datasets. Multiple analysis variables should be separated by a comma.
ANALYSISDATASET	The dataset from which the analysis variables come.
WHERECLAUSEOID	The OID for the where clause that selects proper records for analysis. The WHERECLAUSEOID should exist in the WHERE_CLAUSES spreadsheet.
DOCUMENTATION	A short description of the analysis.
REFLEAFID	The value of a LEAFID in the EXTERNAL_LINKS spreadsheet. Can, for example, point to a section of the SAP that describes the analysis.
CONTEXT	Specifies the software used for the PROGRAMMINGCODE or PROGRAM, if provided.
PROGRAMMINGCODE	In accordance with ADaM principles, ADaM data should be analysis-ready. This would be the place to demonstrate adherence to that principle by providing the code to a statistical procedure that could be used to replicate the results.

Excel Column	Description
PROGRAMLEAFID	The value of a LEAFID in the EXTERNAL_LINKS spreadsheet that points to a script file that can perform the analysis.

The following screen shows a sample of the analysis results metadata produced for our fictional example study data. For readability, the PARAMCD column (column F) has been hidden, and the PROGRAMLEAFID (column N) is not shown.

DISPLAYID	DISPLAYNAME	RESULTNAME	REASON	PURPOSE	ANALYSISVARIABLES	ANALYSISDATASET	WHERECLAUSEOID	DOCUMENTATION	REFLEAFID	CONTEXT	PROGRAMMINGCODE
Table_14.1.1	Summary of Demographics (ITT Population)	Summary of Demographics	SPECIFIED IN SAP	Comparisons of baseline characteristics by treatment group	AGE, AGEGR1, SEX, RACE	ADSL	WC.ITTFL	Rates and chi-squared tests of categorical demographic variables	SAP_Section_9.1.1	SAS Version 9.4	PROC FREQ DATA=ADSL; where ittfl="Y"; tables trt01pn * (agegr1 sex race) / cmh; run;
Table_14.2.1	Responder Analysis by Visit	Responder Analysis by Visit	SPECIFIED IN SAP	Primary efficacy endpoint	CRIT1FL	ADEF	WC.Table_14.2.1.1	Chi-squared test of responder rates by treatment group	SAP_Section_9.2.1	SAS Version 9.4	proc freq data = adef; by crit1 avisitn; where itt="Y"; tables trtpn * crit1 / chisq; run;
Table_14.2.2	Mean Pain Assessments and Changes from Baseline	Summary of Pain Relief	SPECIFIED IN SAP	Secondary efficacy endpoint	AVAL, CHG	ADEF	WC.Table_14.2.1.1	ANCOVA analysis of the treatment group effect with a baseline covariate, by visit	SAP_Section_9.2.1	SAS Version 9.4	proc glm data = adef; by avisitn avisit; class trtpn; model aval = trtpn base; run;
Figure_14.2.1	Time to Pain Relief Analysis by Treatment Group	Time to Pain Relief Kaplan Meier Analysis	SPECIFIED IN SAP	Secondary efficacy endpoint	AVAL	ADTTE	WC.Figure_14.2.1.1	Kaplan-Meier estimates and log-rank test for the treatment effect	SAP_Section_9.2.2	SAS Version 9.4	proc lifetest data = adtte plots=s; where ittfl = "Y" and paramcd = "TTPNRELF"; id usubjid; strata trtpn; time aval*cnsr(1 2 3); test trtpn; run;
Figure_14.2.1	Time to Pain Relief Analysis by Treatment Group	Time to Pain Relief Hazard Ratio Derivation	SPECIFIED IN SAP	Secondary efficacy endpoint	AVAL	ADTTE	WC.Figure_14.2.1.1	Cox proportional hazards model used to derive the hazard ratio for the treatment group effect	SAP_Section_9.2.2	SAS Version 9.4	proc phreg data = adtte; model aval*cnsr(1, 2, 3) = trtpn; run;
Table_14.3.2.1	Incidence Rates of TEAEs by System Organ Class and Preferred Term	Incidence Rates of TEAEs by System Organ Class and Preferred Term	SPECIFIED IN SAP	Secondary safety endpoint	AEBODSYS, AEDECOD	ADAE	WC.Table_14.3.2.1	Incidence rates of TEAEs by System Organ Class and Preferred Term	SAP_Section_9.3.1	SAS Version 9.4	

TOC_METADATA / VARIABLE_METADATA / VALUELEVEL_METADATA / COMPUTATION_METHOD / WHERE_CLAUSES / COMMENTS / CODELISTS / ANALYSIS_RESULTS / EXTERNAL_LINKS

The following screen shows a sample of the external link information needed for the ADaM metadata. Most of the rows are for LEAFIDs specified in the analysis results metadata, either as links to references for analyses (such as sections of the SAP) or to the output display itself, as it appears in, for example, the CSR.

LeafID	LeafRelPath	LeafPageRef	LeafPageRefType	Title	SupplementalDoc	AnnotatedCRF
ADRG	reviewersguide.pdf			Analysis Data Reviewer's Guide	Y	
SAP_Section_9.1.1	..\..\..\..\53-clin-stud-rep\535-rep-effic-safety-stud\5351-stud-rep-contr\studyxyz123\studyxyz123_csr.pdf	SAP_SEC_9.1.1	NamedDestination	SAP Section 9.1.1		
SAP_Section_9.2.1	..\..\..\..\53-clin-stud-rep\535-rep-effic-safety-stud\5351-stud-rep-contr\studyxyz123\studyxyz123_csr.pdf	SAP_SEC_9.2.1	NamedDestination	SAP Section 9.2.1		
SAP_Section_9.2.2	..\..\..\..\53-clin-stud-rep\535-rep-effic-safety-stud\5351-stud-rep-contr\studyxyz123\studyxyz123_csr.pdf	SAP_SEC_9.2.2	NamedDestination	SAP Section 9.2.2		
SAP_Section_9.3.1	..\..\..\..\53-clin-stud-rep\535-rep-effic-safety-stud\5351-stud-rep-contr\studyxyz123\studyxyz123_csr.pdf	SAP_SEC_9.3.1	NamedDestination	SAP Section 9.3.1		
Table_14.1.1	..\..\..\..\53-clin-stud-rep\535-rep-effic-safety-stud\5351-stud-rep-contr\studyxyz123\studyxyz123_csr.pdf	302	PhysicalRef	Table 14.1.1		
Table_14.2.1	..\..\..\..\53-clin-stud-rep\535-rep-effic-safety-stud\5351-stud-rep-contr\studyxyz123\studyxyz123_csr.pdf	352	PhysicalRef	Table 14.2.1		
Figure_14.2.1	..\..\..\..\53-clin-stud-rep\535-rep-effic-safety-stud\5351-stud-rep-contr\studyxyz123\studyxyz123_csr.pdf	353	PhysicalRef	Figure 14.2.1		
ae-t14-3-02	ae-t14-3-02.sas			AE Analysis Program		

Building Define.xml

With all of the metadata groups working together, including ones not shown here but covered in Chapter 2 for the SDTM data, our define.xml file will provide useful links within the file and external links to other documents and files such as data reviewer guides, SAS datasets, output results in the CSR, and original documentation in the SAP or protocol.

The code used to create the SDTM define file using Base SAS is also used for the ADaM define file. You can again refer to the authors' pages (http://support.sas.com/publishing/authors/index.html) for the code and documentation that describe the details of what is being done. Here is an example call of the %MAKE_DEFINE2 macro:

```
%make_define2(path=C:\Projects\StudyXYZ123\data\ADAM-
metadata,metadata=ADAM_METADATA.xlsx);
```

As with the SDTM metadata, all that is needed is to provide the path where the metadata spreadsheet exists and the name of the spreadsheet file itself. The macro will then build the define.xml file in the same directory.

Define.xml Navigation and Rendering

The style sheet provided with the ADaM Analysis Results Metadata release package controls the display of the ADaM metadata used in our example. This particular style sheet has a navigation or bookmark pane on the left side and uses JavaScript to collapse and expand the primary sections into subsections. Such style sheets make navigation of the define file much easier compared to one that does not provide the bookmark pane. Be aware that some browsers or operating system security settings might initially block the JavaScript from running. Even if the JavaScript content is blocked, the style sheet still displays all metadata content, but without the interactive functionality such as folding and unfolding of bookmark pane menu items.

The following screen shows how the define file should look with our metadata and the define2-0-0.xsl style sheet using Firefox v43.0.

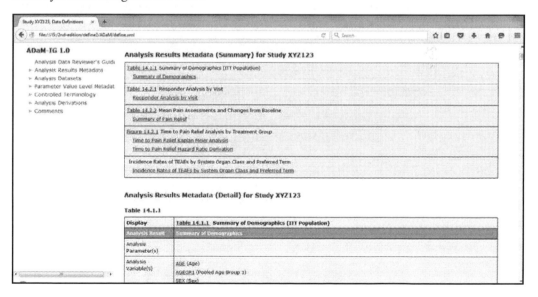

At the top of the define file, you see a table of contents of the analysis results metadata discussed in the previous section. Below that are the details of the analysis results metadata organized by the display ID, and then the analysis result ID within each display. Further down are the data definition tables that display details about each variable from each analysis dataset. The following screen displays these details for ADEF.

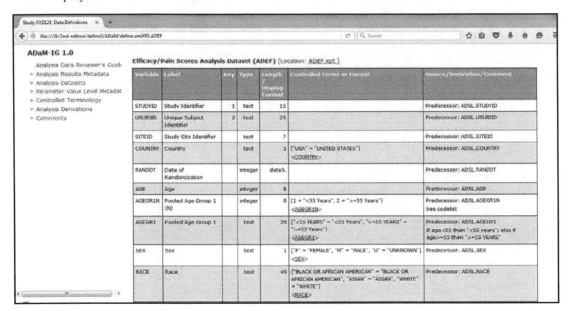

There are too many variables to display all of them in one screenshot. If you could scroll farther down, you could see that the variables AVAL, BASE, CRIT1, and CRIT1FL are in blue and underlined, indicating that they have hyperlinks. These hyperlinks bring you to the analysis parameter value-level metadata. The following screen shows these details. Also shown are the computational methods, including the one used for CRIT1FL.

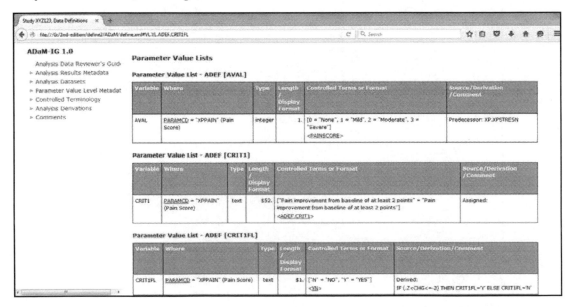

When trying to render or view the define.xml, remember that the style sheet file must exist in the same directory as the define file and the datasets in order for everything to function properly.

Unfortunately, there are many factors that can create rendering problems when you try to view your define file in a browser. Certain combinations of the browser software, browser version, operating system, and security settings can all result in a blank or unreadable display. If you experience such problems in your computing environment, consider using a style sheet that does not use JavaScript, although even this is not guaranteed to resolve all issues.

Chapter Summary

Thanks to the many similarities between SDTM and ADaM metadata, many of the spreadsheets used within our Excel workbook to capture SDTM metadata can also be used to capture ADaM metadata. Likewise, much of the same code used to create the SDTM define file can also be used for the ADaM define file. However, there is one notable piece of metadata that is unique to ADaM: analysis results metadata. In this chapter, we looked closely at the analysis results metadata and also discussed aspects of value-level metadata that are more specific to ADaM.

As mentioned in Chapter 2, having your metadata in place before implementing a CDISC standard will help drive your process and ensure consistency. The same holds true for ADaM metadata, including analysis results metadata. Although certain features, such as links to CSR tables, will not be ready when you are first starting your implementation, certain things will (or can) be ready. Table numbers and titles, documentation, and programming code can be useful metadata components that serve as specifications that your ADaM data might eventually be checked against.

Chapter 7: Implementing ADaM with Base SAS

When your SDTM implementation has been mapped out, your SDTM datasets constructed, and your ADaM metadata is in place, you are ready to start producing your ADaM datasets. In this chapter, we demonstrate an ADaM implementation using an approach similar to the one we used to demonstrate the SDTM conversions in Chapter 3. We produce four ADaM datasets: ADSL, the subject-level dataset; ADEF, a BDS-structured dataset of the pain efficacy data; ADAE, an analysis dataset of the adverse event data; and ADTTE, an analysis dataset for time-to-event data. Keep in mind the two fundamental principles of ADaM, which were also discussed in Chapter 1, as we create these datasets:

1. Keep the data "readily usable by commonly available software tools" (also known as being "one statistical procedure away" or, in SAS parlance, "one PROC away").
2. Make sure that the data provide "traceability between the analysis data and its source data" (ultimately SDTM).

Before diving into this, however, some tools that can be used for common SDTM-to-ADaM conversions will be introduced.

See Also

Analysis Data Model version 2.1 (http://www.cdisc.org/standards/foundational/adam)

ADaM Structure for Occurrence Data (OCCDS) version 1.0
(http://www.cdisc.org/standards/foundational/adam)

ADaM Basic Data Structure for Time-to-Event (TTE) Analyses version 1.0
(http://www.cdisc.org/standards/foundational/adam)

ADaM Tools

As you work on creating ADaM datasets, you will find the need to develop a few tools or macros to assist with automating repetitive tasks such as converting SDTM DTC dates to numeric dates and merging in supplemental qualifier information. These topics are discussed in the following sections.

ISO 8601 Date and DateTime Conversions

Numeric dates in ADaM are typically stored as SAS dates, which represent the number of days since January 1, 1960. Similarly, date/time combination variables are typically stored as SAS date/times, which represent the number of seconds since January 1, 1960. SAS 9.2 has a new class of formats and informats devoted to reading in and creating ISO 8601 dates, times, datetimes, and durations. The informat e8601DA*w.* can be used for converting SDTM DTC date values to SAS dates. The informat i8601DT*w.* is convenient for converting SDTM DTC date/time values to SAS datetimes. The log from the following code demonstrates how these informats can be used.

Log 7.1: Use of ISO 8601 Informats

```
data a;
    adtc = '2008-09-15';
    adt  = input(adtc, e8601da10.);  ❶
    bdtc = adtc || "T15:53:00";
    bdtm = input(bdtc, e8601dt19.);  ❷

    put adt=date9. bdtm=datetime16.;
run;
adt=15SEP2008 bdtm=15SEP08:15:53:00
NOTE: The data set WORK.A has 1 observations and 4
variables.
```

❶ The date width is specified at a length of 10. To prevent errors during execution of your SAS code, you must ensure that the date is complete before trying to convert it to a SAS date. Note that for dates of this length, this informat is no different from YYMMDD10.

❷ As shown, note that the width of the datetime field needs to be 19. Because seconds usually are not captured in clinical trial databases, an SDTM DTC variable with hours and minutes collected would need :00 appended to the field in order to convert the value to a SAS datetime.

Because SDTM DTC values might contain incomplete dates and times, every conversion from a DTC value to a SAS date or datetime must first involve a check of the field's length in order to determine what parts of the value can be converted. A repetitive task such as this, applied to a number of date values used for analysis, would be best handled by a macro. The DTC2DT macro can be used to convert a DTC variable to a SAS date or date/time for an ADaM dataset. The macro can also calculate relative day variables if a reference date is provided as a parameter. This macro is shown below:

```
%macro dtc2dt(dtcvar , prefix=a, refdt= );

    if length(&dtcvar)=10 and index(&dtcvar,'--')=0 then
      &prefix.dt = input(&dtcvar, yymmdd10.);
    else if length(&dtcvar)=16 and index(&dtcvar,'--')=0 and
      index(&dtcvar,'-:')=0 then
```

```
        do;
            &prefix.dtm = input(trim(&dtcvar) ||":00", e8601dt19.);
            &prefix.dt  = datepart(&prefix.dtm);
        end;

    %if &refdt^= %then
    %do;
            if .<&prefix.dt<&refdt then
              &prefix.dy = &prefix.dt - &refdt;
            else if &prefix.dt>=&refdt then
              &prefix.dy = &prefix.dt - &refdt + 1;
    %end;

%mend dtc2dt;
```

First, the macro checks the length of an SDTM DTC variable and then converts the date or datetime to a SAS date or datetime if the length of the field is sufficient. It assumes, by default, that the SAS date or datetime is for an ADaM BDS dataset and the creation of one or a combination of ADT, ADTM, and ADY. (Note, however, that the –DY variable is created only if the REFDT parameter is assigned.) Because the ADaM convention is that all numeric date variables end in DT and that all analysis study day variable names end in DY, these suffixes cannot be changed. However, the prefix can be changed from "A" to something else via the PREFIX parameter.

Let's say you have a lab date in an LB SDTM dataset that you want to convert to ADT for an analysis dataset, ADLB. A sample call of the macro in a DATA step for such a conversion might look like this:

```
Data ADaM.ADLB;
  Merge SDTM.LB ADaM.ADSL (keep=usubjid trtsdt);
    By usubjid;
    .
    .
    .
    ** convert the lab date to a SAS date for variable ADT;
    ** and calculate the study day for variable ADY;
    %dtc2dt(LBDTC, refdt=TRTSDT );
Run;
```

Because the REFDT parameter was assigned, the ADY variable will be created in addition to ADT.

It is common practice to impute values for incomplete dates. What is not common is how this is implemented from one organization to the next or from one study to the next. Imputation conventions can range from the simple, such as always imputing missing days to be 1 (for the first of the month), to the complex. Examples could include imputing one value for events that occur during the same month as the first dose of study medication; another value for events that are known to occur before the first dose; another value for events that are known to occur after the first dose; and then a completely different set of rules depending on whether the date is associated with an efficacy result or a safety result. Whatever your imputation method might be, you can consider incorporating it into the DTC2DT macro.

Merging in Supplemental Qualifiers

You will often need to merge supplemental qualifiers from a SUPP-- SDTM dataset into an ADaM dataset. For example, there might be important dates stored in SUPPDM that will be needed for the creation of ADSL, or a treatment-emergent flag stored in SUPPAE that will be needed for the creation of ADAE.

Consider the sample data in the SUPPDM dataset from Chapter 3. Data from the first four subjects are shown in Table 7.1. All subjects have their date of randomization saved as supplemental qualifiers; and subjects with an Other race have the specification of their race also saved as a supplemental qualifier. Both pieces of information might be needed for construction of the ADSL dataset. To do this, you would have to first transpose SUPPDM, converting records where QNAM='RACEOTH' and QNAM='RANDDTC' into a horizontal structure with RACEOTH and RANDDTC as variables, and then merge those variables in with the ADSL dataset.

Table 7.1: Supplemental Qualifiers to DM

STUDYID	RDOMAIN	USUBJID	QNAM	QLABEL	QVAL	QORIG
XYZ123	DM	UNI101	RACEOTH	Race, Other	LAOTIAN	CRF Page 1
XYZ123	DM	UNI101	RANDDTC	Date of Randomization	2010-04-02	CRF Page 1
XYZ123	DM	UNI102	RANDDTC	Date of Randomization	2010-02-13	CRF Page 1
XYZ123	DM	UNI103	RANDDTC	Date of Randomization	2010-05-16	CRF Page 1
XYZ123	DM	UNI104	RANDDTC	Date of Randomization	2010-01-02	CRF Page 1

The %MERGSUPP macro was designed for this task. It can be used either to merge supplemental qualifiers from a specified list of domains or, if the DOMAINS parameter is left blank, to merge all supplemental qualifiers with all parent domains in a given source library. This macro is below, followed by an explanation of certain points along the way.

```
*------------------------------------------------------------;
* Merge supplemental qualifiers into the parent SDTM domain  ;
* This can either be for an entire library or for specified   ;
* domains                                                     ;
*------------------------------------------------------------;
%macro mergsupp(sourcelib=library, outlib=WORK, domains= , suppqual=0);  ❶

  %local domain;

  %** de-normalize suppqual and merge into the given domain;
  %macro domainx(domain= ,suppqual=0);

    %local suppdata idvar varlist nvars;

    %if &suppqual %then
      %let suppdata=suppqual;
    %else
      %let suppdata=supp&domain;
    ;

    %* count the number of supplemental qualifiers for the given
      domain;
     proc sort
      data = &sourcelib..&suppdata
      out = nvars
      nodupkey;    ❷
        where rdomain=upcase("&domain");
        by qnam idvar;
    run;

    data _null_;
      set nvars end=eof;
        by qnam idvar;

          length varlist $200;
          retain varlist;
          if not first.qnam then
           put 'PROB' 'LEM: More than one IDVAR for the domain-- '
                  rdomain= qnam=idvar= ;
          else
             do;
                 nvars + 1;
                 varlist = trim(varlist) || " " || trim(qnam);
             end;
```

```
        if eof then
          do;
              call symput("nvars", put(nvars, 2.));
              call symput("varlist", trim(left(varlist)));
              call symput("idvar", trim(idvar));
          end;
run;
%put domain=&domain idvar=&idvar nvars=&nvars varlist=&varlist;  ❸


proc sort
  data = &sourcelib..&suppdata
  out = supp&domain;
     where rdomain=upcase("&domain");
     by usubjid idvar idvarval;
run;


%*  determine whether IDVAR in the parent domain is character or
    numeric;  ❹
%if &idvar^= %then
  %do;
     %let dsetnum=%sysfunc(open(&sourcelib..&domain));
     %let varnum=%sysfunc(varnum(&dsetnum,&idvar));
     %let idtype=%sysfunc(vartype(&dsetnum,&varnum));
     %let rc=%sysfunc(close(&dsetnum));
  %end;
%else
     %let idtype= ;


data supp&domain;
  set supp&domain;
    by usubjid idvar idvarval;

       drop qnam qval idvarval idvar i rdomain;
       length &varlist $200.;
       retain &varlist;
       array vars{*} &varlist;
       if first.idvarval then
         do i = 1 to dim(vars);
            vars{i} = '';
         end;

       do i = 1 to dim(vars);
         if upcase(qnam)=upcase(vname(vars{i})) then
            vars{i} = qval;
       end;

       %** convert to numeric if numeric in the parent domain;
       %if &idvar^= and &idtype=N %then
          &idvar = input(idvarval, best.);
       %else %if &idvar^= %then
          &idvar = idvarval;
          ;
          if last.idvarval;
run;


proc sort
  data = supp&domain;
    by usubjid &idvar;
```

```
proc sort
  data = &sourcelib..&domain
  out = __tmp;
    by usubjid &idvar;

data &outlib..&domain;
  merge __tmp supp&domain ;  ❺
    by usubjid &idvar;
run;

%mend domainx;

%*----------------------------------------------------------;
%* If DOMAINS parameter specified, then loop through those ;
%* domains otherwise, dynamically identify the SUPPxx data ;
%* sets and go through them all                           ;
%*----------------------------------------------------------;
%let _wrd=1;
%if &DOMAINS^= %then  ❻
  %do %while(%scan(&domains,&_wrd)^= );
      %let domain=%scan(&domains,&_wrd);
      %domainx(domain=&domain,suppqual=0);
      %let _wrd=%eval(&_wrd+1);
  %end;
%else
  %do;
      %** find all of the SUPPxx datasets and loop through each one;
      ods output members=members;
      proc contents
        data = &sourcelib.._all_ memtype=data nods ;
      run;

      data members;
        set members;

          if upcase(name)=:'SUPP' and upcase(name)^=:'SUPPQUAL' then
                  do;
                  rdomain = substr(name,5,2);
                  put name= rdomain= ;
                  output;
                end;
              else if upcase(name)=:'SUPPQUAL' then
                call symput("suppqual","1");
  run;

  %** loop through each domain;
  proc sql noprint;
    select count(distinct rdomain)
        into :domn
        from %if &suppqual %then &sourcelib..suppqual; %else
            members;
        ;
    select distinct rdomain
        into :domain1 - :domain%left(&domn)
        from %if &suppqual %then &sourcelib..suppqual; %else
            members;
        ;
    %do _i=1 %to &domn;
      %domainx(domain=&&domain&_i,suppqual=&suppqual);
    %end;
```

```
            %end; %* if domains not specified explicitly...;

%mend mergsupp;
```

❶ In the macro call, the SOURCELIB parameter is used to specify the LIBNAME where all of the SDTM datasets exist. The OUTLIB parameter defaults to WORK, so that the resulting domain or domains with the supplemental qualifiers are created as WORK datasets. The DOMAINS parameter can be used to explicitly define domains for which you want to merge in any supplemental qualifiers.

❷ For each domain, the unique number of supplemental qualifiers, which will become variables, is determined.

❸ The domain, ID variable for the domain (which is used as a part of the unique key for the dataset), number of supplemental qualifiers, and the list of unique QNAM values are recorded in the SAS log.

❹ When merging with the parent domain, the variable defined by IDVAR must match in type with the parent domain. Otherwise, an error will occur. Because the value of the IDVAR can be stored only as a character in the SUPP dataset, it must be converted to numeric when the data are transposed if the IDVAR is numeric in the parent domain.

❺ In this step, the supplemental qualifiers are finally merged with the parent domain.

❻ If the DOMAINS parameter is not blank, then this routine is performed for each domain in the DOMAINS list. If the DOMAINS parameter is blank, then the following conditions are checked: If SUPPQUAL=1, then the routine is run for each domain found in the SUPPQUAL dataset. Otherwise, the source library is scoured for datasets that begin with "SUPP".

The following sections demonstrate how these tools can be used for the creation of ADaM datasets.

ADSL – The Subject-Level Dataset

One of two dataset structures described in the ADaM IG is ADSL, the subject-level analysis dataset. (ADSL is both the name of the structure and the expected name of the dataset that follows that structure.) The existence of an ADSL dataset is the minimum requirement for an ADaM submission. It is extremely important as a source for key information about every randomized or treated subject in a clinical trial. It contains key demographic data, treatment information, and population flags. It can often contain additional information such as final study disposition; safety information (for example, a death flag); and sometimes even key outcome data (for example, whether the subject was a responder).

As you can see, a lot of data are merged together when creating ADSL. As mentioned in the previous section, sometimes these data must come from supplemental qualifiers. The following code demonstrates some of the common conversions and variable declarations that go into an ADSL creation program.

```
    *-----------------------------------------------------------*;
    * ADSL.sas creates the ADaM ADSL data set
    * as permanent SAS datasets to the ADaM libref.
    *-----------------------------------------------------------*;

    **** CREATE EMPTY ADSL DATASET CALLED EMPTY_ADSL;
    %let metadatafile=&path/data/adam-metadata/adam_metadata.xlsx
    %make_empty_dataset(metadatafile=&metadatafile, dataset=ADSL) ❶
```

```
** merge supplemental qualifiers into DM;
%mergsupp(sourcelib=sdtm, domains=DM);  ❷

** find the change from baseline so that responders can be flagged;
** (2-point improvement in pain at 6 months);
%cfb(indata=sdtm.xp, outdata=responders, dayvar=xpdy, avalvar= xpstresn,
     keepvars=usubjid visitnum chg);  ❸

data ADSL;
    merge EMPTY_ADSL
              DM             (in = inDM)
              responders  (in = inresp where=(visitnum=2))
              ;
        by usubjid;

        * convert RFXSTDTC to a numeric SAS date named TRTSDT;
        %dtc2dt(RFXSTDTC, prefix=TRTS );   ❹

        * create BRTHDT, RANDDT, TRTEDT;
        %dtc2dt(BRTHDTC, prefix=BRTH);
        %dtc2dt(RANDDTC, prefix=RAND);
        %dtc2dt(RFXENDTC, prefix=TRTE);

        * created flags for ITT and safety-evaluable;
        ittfl = put(randdt, popfl.);
        saffl = put(trtsdt, popfl.);

        trt01p = ARM;
        trt01a = trt01p;
        trt01pn = input(put(trt01p, $trt01pn.), best.);
        trt01an = trt01pn;   ❺

        agegr1n = input(put(age, agegr1n.), best.);
        agegr1  = put(agegr1n, agegr1_.);

        RESPFL = put((.z <= chg <= -2), _0nly.);           ❻
run;

**** SORT ADSL ACCORDING TO METADATA AND SAVE PERMANENT DATASET;
%make_sort_order(metadatafile=&metadatafile, dataset=ADSL)  ❼

proc sort
  data=adsl
  (keep = &ADSLKEEPSTRING)
  out=adam.adsl;
    by &ADSLSORTSTRING;
run;
```

❶ Similar to what was done for SDTM conversions, the %MAKE_EMPTY_DATASET macro is used to create a 0-observation dataset from the metadata.

❷ As demonstrated in the previous section, the %MERGSUPP macro can be used to merge in supplemental qualifiers from the SDTM data that might be needed either as columns in the ADaM data or for deriving columns in the ADaM data. In this case, the randomization date is used for both—as a column in ADSL and for calculating relative days in ADaM. In this example, relative days are defined by using the randomization date as the anchor date rather than the date of first dose.

❸ Sometimes it is necessary to have efficacy results, such as a flag for study responders, in ADSL. However, this often creates circular references if the code to derive the efficacy result first relies on the existence of ADSL. A way around this, and to avoid having duplicated code, is to create a macro that does the derivation and is called by multiple programs—in this case, ADSL.SAS and the program devoted to efficacy results at all visits (ADEF.SAS, which is shown later). Both programs call the %CFB macro (shown later) to derive changes from baseline so that responders can be easily identified.

❹ As shown in the previous section, the %DTC2DT macro is used to convert SDTM --DTC dates to numeric (SAS) dates. In this particular spot, the macro is used to create TRTSDT, the date of first treatment.

❺ In this example, we are assuming that all subjects took the dose to which they were randomized, so TRT01A is set equal to TRT01P. However, in many real-life situations, this is not the case; and this assignment is not, therefore, always valid. Because actual treatment information is not known at the beginning of a trial when dataset specifications are developed, the convention applied here is to include TRT01A and TRT01AN regardless of whether they differ from the planned assignment. This also allows you to begin table programming without needing to know the actual treatment results (if that approach is desired).

❻ Here the CHG variable created within the %CFB macro is used to create the responder flag.

❼ As was done with the SDTM conversion programs, the %MAKE_SORT_ORDER macro is used to properly sort the data. The &ADSLKEEPSTRING macro variable, which was created by the %MAKE_EMPTY_DATASET macro, is used to keep only those variables in the metadata.

The following code shows the %CFB macro. As stated, this macro is also used for the efficacy analysis dataset and is discussed more in the next section. Its primary purpose is to calculate changes from baseline (variable CHG). By doing so, it also creates common analysis variables AVAL and PCHG (percent change from baseline), the baseline value (BASE), and a flag for the baseline record (ABLFL). For ADSL, the CHG variable is needed to derive the responder flag (RESPFL). In the next section, we will see it used for an analysis dataset with a different ADaM structure.

```
*-------------------------------------------------------------;
* Change From Baseline                                        ;
* Macro for deriving ABLFL, BASE, CHG, and PCHG for a BDS     ;
*    formatted ADaM data set;                                 ;
* Assumes baseline is the last non-missing value on or before ;
*    study day 1 and that the INDDATA is an SDTM data set with ;
*    variables USUBJID and VISITNUM                           ;
*-------------------------------------------------------------;
%macro cfb(indata= ,outdata= ,avalvar= ,dayvar= ,keepvars= );

    proc sort
      data = &indata
      out = &outdata (rename = (&avalvar = aval));
        by usubjid visitnum;
    run;

    * Baseline is defined as the last non-missing value prior to study
      day 1 first dose;
    * (note, values on Day 1 are assumed to occur before the first dose);
    data base1 (keep = usubjid visitnum) base2 (keep = usubjid base);
      set &outdata;
        where &dayvar<=1 and aval > .z;
```

```
        by usubjid visitnum;

      rename aval = base;
      if last.usubjid;
  run;


  * Do one merge to identify the baseline record;
  data &outdata;
     merge &outdata base1 (in = inbase);
         by usubjid visitnum;

         if inbase then
            ablfl = 'Y';
  run;


  * Do another merge to get the baseline value;
  data &outdata;
     merge &outdata base2;
         by usubjid;

         %if &keepvars^= %then
           keep   &keepvars;
         ;

         if base > .z then
           chg  = aval - base;
         if base>.z and base ne 0 then
           pchg = chg/base*100;
  run;


%mend cfb;
```

Note that the typical needs of a real-life study can be much more complex than what is being shown here. For the sake of simplicity, methods for dealing with missing values, visits that fall outside of a particular visit window, or unscheduled visit values are not dealt with here.

The ADaM Basic Data Structure (BDS)

The ADaM Basic Data Structure (BDS) is the other of the two data structures detailed in the ADaM IG. Unlike ADSL, which is specific to one intended dataset, the BDS is a general structure—flexible enough to use for a number of different analysis datasets. It can loosely be described as a "tall and skinny" structure where, for example, each row contains one result per test, per analysis visit, per subject.

In this section, we demonstrate an ADaM BDS dataset for the creation of an efficacy dataset (ADEF) using the XP (pain) SDTM data shown in earlier chapters. Consider the structure of this dataset shown in Table 7.2. It contains one record per visit, per subject. In this simplified case there is only one test or result per visit. In a more complex (and probably more typical) arrangement, more than one related efficacy result would be collected at each visit. If this were the case, the SDTM data would contain a new row with a new test and test code (that is, XPTEST and XPTESTCD).

Table 7.2: Snippet of XP SDTM Data

USUBJID	XPSEQ	XPTESTCD	XPTEST	XPORRES	XPSTRESN	VISIT	XPDTC	XPDY
UNI101	1	XPPAIN	Pain Score	Severe	3	Baseline	2010-04-02	1
UNI101	2	XPPAIN	Pain Score	Moderate	2	3 Months	2010-07-03	93
UNI101	3	XPPAIN	Pain Score	Mild	1	6 Months	2010-10-10	192

Fortunately, for implementers of ADaM, the BDS structure is similar to SDTM domains of the Findings class (although the BDS structure is by no means limited to findings). The conversion to ADaM in this simplified example therefore involves little more than keeping, dropping, renaming, and adding a few fields. No complicated transformations are needed. This is demonstrated in the following code. The primary fields being added are the BASE and CHG variables, which are needed for the efficacy analysis.

```
*-------------------------------------------------------------*;
* ADEF.sas creates the ADaM BDS-structured data set           *;
* for efficacy data (ADEF), saved to the ADaM libref.         *;
*-------------------------------------------------------------*;

**** CREATE EMPTY ADSL DATASET CALLED EMPTY_ADSL;
%let metadatafile=&path/data/adam-metadata/adam_metadata.xlsx
%make_empty_dataset(metadatafile=&metadatafile,dataset=ADEF)

** derive AVAL, BASE, CHG, and PCHG;
%cfb(indata=sdtm.xp, outdata=adef, dayvar=xpdy, avalvar= xpstresn);  ❶

proc sort
  data = adam.adsl
  (keep = usubjid siteid country age agegr1 agegr1n sex race randdt trt01p
trt01pn ittfl)
  out = adsl;
    by usubjid;

data adef;
  merge adef (in = inadef) adsl (in = inadsl);
    by usubjid ;

        if not(inadsl and inadef) then
          put 'PROB' 'LEM: Missing subject?-- ' usubjid= inadef= inadsl= ;

        rename trt01p    = trtp   ❷
               trt01pn   = trtpn
               xptest    = param
               xptestcd  = paramcd
               visit     = avisit
               visitnum  = avisitn
               xporres   = avalc
        ;
        if inadsl and inadef;
        %dtc2dt(xpdtc, refdt=randdt);    ❸

        retain crit1 "Pain improvement from baseline of at least 2 points";
        crit1fl = put((.z <= chg <= -2), _Only.);   ❹
  run;

** assign variable order and labels;
data adef;
```

```
    retain &ADEFKEEPSTRING;
    set EMPTY_ADEF adef;
run;


**** SORT ADEF ACCORDING TO METADATA AND SAVE PERMANENT DATASET;
%make_sort_order(metadatafile=&metadatafile, dataset=ADEF)

proc sort
    data=adef(keep = &ADEFKEEPSTRING)
    out=adam.adef;
      by &ADEFSORTSTRING;
    run;
```

❶ Here the %CFB macro used in ADSL is again used for the creation of ADEF. The difference now is that all visits from the resulting WORK.ADEF dataset are kept, as are the variables AVAL, BASE, CHG, and ABLFL (via the metadata, the %make_empty_dataset macro, and the &ADEFKEEPSTRING macro variable).

❷ As mentioned, when converting an SDTM findings domain to the BDS, the structure is essentially the same. But many variable names change, and the crucial analysis variables such as AVAL and CHG derived by the %CFB macro have been added. Although using the RENAME statement can potentially apply incorrect labels and lengths to the new variables, the consequences are small because the next DATA step uses the EMPTY_ADEF dataset to assign proper labels and lengths.

❸ Here the %DTC2DT macro is used to create both ADT and, by virtue of specifying a REFDT, the relative day.

❹ The code here to define responders is similar to that used in ADSL. The primary difference is that here we are defining responders at the visit level. But in ADSL, we were defining study-wide responders (those who met the criteria at the primary time point or visit). The variable name, CRIT1FL, is different as well, in order to use BDS naming conventions. (We also could have used an AVALCATy variable such as AVALCAT1.)

As shown in the previous section, the %CFB macro assigns pain scores from the SDTM data to AVAL and calculates parameter-invariant fields BASE and CHG. (PCHG is also created but not used for this dataset.) *Parameter invariant* means that the derivation does not change from one parameter to the next. In this simplified example, there is only one PARAM in ADEF. Generally speaking, it would be best to derive all parameter-invariant columns added to a BDS dataset, such as BASE or CHG, in one DATA step for all parameters at once. The advantage of doing this is that it ensures that the same algorithm was applied for all parameters and that the code for that algorithm is not repeated.

ADAE – Adverse Event Analysis Datasets

As a follow-up to the ADaM team's ADAE document, an entirely new ADaM data structure has been established— the Occurrence Data Structure, or OCCDS. ADAE can now be considered an example of the OCCDS data structure. The structure of ADAE is very similar to that of the SDTM AE dataset. The advantage of ADAE is that it includes other variables used for analysis, such as population flags, treatment variables, and imputations of severity for cases where the severity is missing. Other variables that are useful for AE analyses include those that flag events within a certain pool of events. Assuming that AEs have been coded to the MedDRA dictionary, pooled events can either be grouped based on standardized MedDRA Queries (SMQs) or on sponsor-defined customized queries (CQs).

The following code demonstrates the creation of an ADAE file using the AE SDTM dataset first created in Chapter 3. It uses the variable CQ01NAM to identify events relating to pain (which are important to know about in a pain study).

```
*----------------------------------------------------------*;
* ADAE.sas creates the ADaM ADAE-structured data set       *;
* for AE data (ADAE), saved to the ADaM libref.            *;
*----------------------------------------------------------*;

**** CREATE EMPTY ADAE DATASET CALLED EMPTY_ADAE;
%let metadatafile=&path/data/adam-metadata/adam_metadata.xlsx
%make_empty_dataset(metadatafile=&metadatafile,dataset=ADAE)

proc sort
  data = adam.adsl
  (keep = usubjid siteid country age agegr1 agegr1n sex race trtsdt trt01a
trt01an saffl)
  out = adsl;
    by usubjid;

data adae;
  merge sdtm.ae (in = inae) adsl (in = inadsl);
    by usubjid ;

        if inae and not inadsl then
          put 'PROB' 'LEM: Subject missing from ADSL?-- ' usubjid= inae=
              inadsl= ;

        length CQ01NAM $40.;
        rename trt01a  = trta    ❶
               trt01an = trtan
        ;
        if inadsl and inae;

        %dtc2dt(aestdtc, prefix=ast, refdt=trtsdt);    ❷
        %dtc2dt(aeendtc, prefix=aen, refdt=trtsdt);

        if index(upcase(AEDECOD), 'PAIN')>0 or upcase(AEDECOD)='HEADACHE' then  ❸
          CQ01NAM = 'PAIN EVENT';
        else
          CQ01NAM = ' ';

        aereln = input(put(aerel, $aereln.), best.);
        aesevn = input(put(aesev, $aesevn.), best.);
        relgr1n = (aereln>0); ** group related events (AERELN>0);
        relgr1  = put(relgr1n, relgr1n.);

        * Event considered treatment emergent if it started on or after;
        * the treatment start date.  Assume treatment emergent if start;
        * date is missing (and the end date is either also missing or >=;
        * the treatment start date)  ;
          trtemfl = put((astdt>=trtsdt or (astdt<=.z and
                    not(.z<aendt<trtsdt))), _0n1y.);
run;

** assign variable order and labels;
data adae;
  retain &adaeKEEPSTRING;
  set EMPTY_adae adae;
run;
```

```
**** SORT adae ACCORDING TO METADATA AND SAVE PERMANENT DATASET;
%make_sort_order(metadatafile=&metadatafile, dataset=ADAE)

proc sort
  data=adae(keep = &adaeKEEPSTRING)
  out=adam.adae;
    by &adaeSORTSTRING;
run;
```

❶   Contrary to efficacy analyses, safety analyses are usually conducted with treatment group assignments based on the drug subjects actually received, rather than on what they were randomized to, hence the use of TRT01A and TRT01AN from ADSL. Although there is usually no difference between what a subject actually took and what a subject was supposed to take, it is helpful for planning and for pre-programming purposes (that is, being able to program before a study is unblinded) to have variables that can differentiate between the two.

❷   Other differences from efficacy analyses are the population studied and the calculation of relative day. Safety analyses are usually based on a population represented by all subjects who received the study drug, rather than by all subjects randomized. Consequently, relative days are calculated as the number of days since first dose rather than the number of days since the subject was randomized. (Again, in most cases you would expect these two values to be the same.) The REFDT in the %DTC2DT macro is therefore TRTSDT rather than RANDDT used for ADEF.

❸   CQ01NAM is used to identify adverse events related to occurrences of pain. Rather than having a flag variable, these are simply signified by populating CQ01NAM with the name for this customized MedDRA query. A CQ would typically be a bit more thorough than simply including any term with the word "PAIN" or "HEADACHE" and would typically be defined before knowing the precise terms or MedDRA codes that appear in a study. But for the purposes of illustration, this example should suffice.

ADTTE – The Time-to-Event Analysis Dataset

Now that we have our ADEF and ADAE datasets created, we can illustrate the creation of an ADaM dataset for time-to-event analyses. Suppose the SAP includes an endpoint for time to first pain relief without pain worsening. Events are defined by records in ADEF where the change in pain is negative. Subjects with pain that gets worse before it improves are censored at the time that the pain gets worse. The censored records will therefore come from records in ADEF where the change in pain is positive. Suppose the SAP states that adverse event data are also considered for identifying subjects with pain that gets worse before it improves. This is where our customized query comes in handy. Subjects with a nonmissing CQ01NAM that occurs before the first pain relief or before the first pain worsening will be censored at the time of the pain-related AE.

The ADaM Basic Data Structure for Time-to-Event Analyses document (http://www.cdisc.org/standards/foundational/adam) demonstrates precisely how you should comply with the ADaM principle of traceability by reflecting in your data how the derivations for such an analysis were implemented. The EVNTDESC variable is used to describe what event occurred. For this example, we will define four possible values, one for an actual event of pain relief, and three for censoring events. A unique value for CNSR will be assigned for each of these events, as shown in Table 7.3.

Table 7.3: CNSR and EVNTDESC Values

EVNTDESC	CNSR	Comment
PAIN RELIEF	0	Actual events of pain relief that occur prior to pain worsening
PAIN WORSENING PRIOR TO RELIEF	1	Pain worsening from ADEF
PAIN ADVERSE EVENT PRIOR TO RELIEF	2	Pain AE prior to pain relief
COMPLETED STUDY PRIOR TO RELIEF	3	Standard censoring at study end

With this plan in mind, the following code can be used to create ADTTE from ADEF, ADAE, and ADSL.

```
*-------------------------------------------------------------*;
* ADTTE.sas creates the ADaM BDS-structured data set for a    *;
* time-to-event analysis (ADTTE), saved to the ADaM libref.   *;
*-------------------------------------------------------------*;

**** CREATE EMPTY ADTTE DATASET CALLED EMPTY_ADTTE;
options mprint ;
%let metadatafile=&path/data/adam-metadata/adam_metadata.xlsx
%make_empty_dataset(metadatafile=&metadatafile,dataset=ADTTE)

proc sort
  data = adam.adsl
  (keep = studyid usubjid siteid country age agegr1 agegr1n sex race
randdt trt01p trt01pn ittfl trtedt)
  out = adtte;
    by usubjid;

proc sort
  data = adam.adef
  (keep = usubjid paramcd chg adt visitnum xpseq)
  out = adef;
    where paramcd='XPPAIN' and visitnum>0 and abs(chg)>0;
    by usubjid adt;

data adef;
  set adef;
    by usubjid adt;

        drop paramcd visitnum;
        if first.usubjid; ❶
run;

proc sort
  data = adam.adae
  (keep = usubjid CQ01NAM astdt trtemfl aeseq)
  out = adae;
```

```
      where CQ01NAM ne '' and trtemfl='Y';
      by usubjid astdt;
run;


** keep only the first occurence of a pain event;
data adae;
  set adae;
    by usubjid aesdt;

        if first.usubjid; ❷
run;


data adtte;
  merge adtte (in = inadtte rename=(randdt=startdt))
        adef  (in = inadef)
        adae  (in = inadae)
        ;
    by usubjid ;

        retain param "Time to first pain relief (days)"
               paramcd "TTPNRELF"
        ;
        rename trt01p    = trtp
               trt01pn   = trtpn
        ;

        length srcvar $10. srcdom $4.;

        if (.<chg<0) and (adt<astdt or not inadae) then   ❸
          do;
            ** ACTUAL PAIN RELIEF BEFORE WORSENING;
            cnsr = 0;
            adt  = adt;
            evntdesc = put(cnsr, evntdesc.) ;
            srcdom = 'ADEF';
            srcvar = 'ADY';
            srcseq = xpseq;
          end;
        else if chg>0 and (adt<astdt or not inadae) then
          do;
            ** CENSOR: PAIN WORSENING BEFORE RELIEF;
            cnsr = 1;
            adt  = adt;
            evntdesc = put(cnsr, evntdesc.) ;
            srcdom = 'ADEF';
            srcvar = 'ADY';
            srcseq = xpseq;
          end;
        else if (.<astdt<adt) then
          do;
            ** CENSOR: PAIN AE BEFORE RELIEF;
            cnsr = 2;
            adt  = astdt;
            evntdesc = put(cnsr, evntdesc.) ;
            srcdom = 'ADAE';
```

```
              srcvar = 'ASTDY';
              srcseq = aeseq;
          end;
      else
          do;
            ** CENSOR: COMPLETED STUDY BEFORE PAIN RELIEF OR WORSENING;
            cnsr = 3;
            adt  = trtedt;
            evntdesc = put(cnsr, evntdesc.) ;
            srcdom = 'ADSL';
            srcvar = 'TRTEDT';
            srcseq = .;
          end;

      aval = adt - startdt + 1;

      format adt yymmdd10.;
run;


** assign variable order and labels;
data adtte;
  retain &adtteKEEPSTRING;
  set EMPTY_adtte adtte;
run;

**** SORT adtte ACCORDING TO METADATA AND SAVE PERMANENT DATASET;
%make_sort_order(metadatafile=&metadatafile, dataset=ADTTE)

proc sort
  data=adtte(keep = &adtteKEEPSTRING)
  out=adam.adtte;
    by &adtteSORTSTRING;
run;
```

❶ From ADEF, the first worsening or pain improvement event is saved and merged with ADSL and ADAE.

❷ From ADAE, the first treatment-emergent, pain-related AE is kept and merged with ADSL and ADEF.

❸ Condition-based code exists for each censoring reason. Note the use of the SRC--- variables for traceability back to other ADaM datasets. The ADaM IG mentions that "traceability is built by clearly establishing the path between an element and its immediate predecessor." As such, despite the reserved name of SRCDOM, which implies that the source is an SDTM domain, the referenced source is actually another ADaM dataset. Note also, however, that the SRCSEQ variable is still from the original SDTM source that was carried over to the ADaM dataset. For this example, this is acceptable because the SRCSEQ variables still uniquely identify subjects' data records. This will not always be the case because one record from an SDTM source file can sometimes be used for multiple ADaM records (for example, when a Last Observation Carried Forward algorithm is implemented). In situations where the SDTM --SEQ variable does *not* uniquely identify an ADaM record, the variable ASEQ can be added to the ADaM data for traceability purposes.

Chapter Summary

Analysis datasets are a perfect complement to the SDTM. While the SDTM serves the purpose of providing all data as collected (at least those data that ultimately could be used, in some way, to evaluate a product's safety or efficacy), ADaM provides analysis-ready data that can quickly be used for producing (or re-producing) important study results.

In this chapter, we illustrated the implementation of ADaM using fairly straightforward, though somewhat common, examples. For some of the common ADaM creation tasks, such as converting SDTM --DTC dates to SAS dates and merging supplemental qualifiers into a parent domain, macros have been provided for standardizing the process. Many other macros introduced during the Base SAS SDTM implementation in Chapter 3 were also applied here, thus illustrating that the advantages of setting up your specifications and metadata ahead of time apply to the ADaM creation process as well.

The true complexities with creating ADaM data tend to fall with the study-specific derivations themselves. Derivations that are somewhat common across studies, such as a Last Observation Carried Forward for missing value imputations (despite its many criticisms it is still widely applied), or assigning analysis visit values based on a windowing algorithm, could be added to your ADaM implementation toolbox, applying the rules generally used for such algorithms within your organization.

Chapter 8: CDISC Validation Using SAS

As we begin to discuss CDISC validation, it might make sense to first clarify what is meant by the term *validation*. Two types of checks are important in this context: 1) ensuring that the data contents are, in fact, what they should be and 2) ensuring that the file structures are what they should be. The latter is the focus of this chapter and other parts of this book, although checks relating to the appropriate use of controlled terminology, which are also covered here, can be construed as the former type.

Human intervention is needed most for the first type of validation check, where ensuring that the collected data values have been correctly converted or mapped to the SDTM. This conversion process, for the most part, is unique to every trial. Standardized file and data structures, on the other hand, vary little from one project to the next. These types of checks, which we also refer to as compliance checks, lend themselves well to the use of automated tools. At the time of this writing, there are a handful of tools available to CDISC implementers for checking CDISC compliance. The SAS Clinical Standards Toolkit, SAS Clinical Data Integration (which uses the toolkit), and the Pinnacle 21Community Validator (previously known as OpenCDISC) are three such compliance tools.

There is another important thing to consider before your first attempt at CDISC data validation: Achieving 100% compliance (according to the tool being used) tends to be an unattainable goal. Too often there are issues either with the collected data, the study design, the implementation strategy, or the compliance checks themselves that prevent you from making all errors and warnings go away. Because of this, when you are submitting data to the FDA or to anyone involved with inspecting the data, it might be a good idea to point out certain checks that fail and the reason why they cannot be corrected. These issues can be documented in the Study Data Reviewer's Guide or the Analysis Data Reviewer's Guide.

SAS Clinical Standards Toolkit SDTM Validation

The SAS Clinical Standards Toolkit that we explored in Chapter 4 for the production of SDTM data sets can also be used to validate SDTM data sets. We will build on the metadata introduced in Chapter 4 to use the SAS Clinical Standards Toolkit to validate those SDTM data sets created in Base SAS using the SAS Clinical Standards Toolkit. This example is built on information provided in the *SAS Clinical Standards Toolkit 1.6: User's Guide.*

SAS Clinical Standards Toolkit Setup

SAS Clinical Standards Toolkit jobs are governed by various metadata data sets, control data sets, a properties file, and finally a SAS program that pulls it all together. To validate the SDTM data sets that were created in Chapter 4, we need a validation checks control data set; a study-specific controlled terminology format catalog; a set of study and target standard metadata data sets; a configuration validation properties file; and a sasreferences data set to tell the final validation SAS program where everything can be found for the job. The first step that you should take is to create an area in which to perform the SDTM validation. For this exercise, that is C:\Users\yourid\Desktop\SAS_BOOK_cdisc_2nd_edition\CST\validate_sdtm. You can replace that with any directory that you would like to use as a default so long as you change it in the following examples as well. For this chapter, the following subdirectories should be created under \validate_sdtm: \control, \data, \metadata, \programs, \results, and \terminology.

Validation Control Data Set

The SAS Clinical Standards Toolkit is preloaded with three sets of checks, including the Oracle Health Sciences WebSDM, OpenCDISC (discussed in a later chapter, and now called Pinnacle 21 Community), and checks that are based on SAS, along with the SDTM validation checks for multiple versions of the SDTM. Please note in SAS Clinical Standards Toolkit 1.7 that the OpenCDISC/Pinnacle 21 checks have now been removed from the toolkit. All of the checks can be found in the SAS toolkit global library for your standard, which is provided in the validation_master.sas7bdat data set under the \validation\control folder. However, you might not want to institute every provided check. You should create a validation_control.sas7bdat control data set based on the validation_master.sas7bdat data set that contains just the SDTM checks that you want to run. For this exercise, we will run the OpenCDISC and WebSDM checks for SDTM version 3.2 only. The following program creates that data set:

```
libname master "C:\cstGlobalLibrary\standards\XYZ_program_cdisc-
sdtm-3.2-1.6\validation\control";
libname control "C:\Users\yourid\Desktop\
SAS_BOOK_cdisc_2nd_edition\CST\validate_sdtm\control";

data control.validation_control;
  set master.validation_master;

  where standardversion in ("***","3.2")
        and checksource in ("OpenCDISC","WebSDM");
run;
```

This program takes a copy of the core XYZ-SDTM-3.2 standard master validation data set called validation_master.sas7bdat and subsets it just for the checks that we want into a data set called validation_control.sas7bdat.

Study-Based Custom Codelist Format Catalog

In Chapter 4, we stored our custom formats in a format.sas7bcat file that contained our customized controlled terminology for the XYZ123 study. You will want to do the same thing here and store that data under
C:\Users\yourid\Desktop\SAS_BOOK_cdisc_2nd_edition\CST\validate_sdtm\terminology\formats. Here is what the control file looks like for that format catalog:

	fmtname	start	end	label	type	hlo
1	ARM	Analgezia HCL 30	Analgezia HCL 30	Analgezia HCL 30 mg	C	
2	ARM	Placebo	Placebo	Placebo	C	
3	ARMCD	ALG123	ALG123	ALG123	C	
4	ARMCD	PLACEBO	PLACEBO	PLACEBO	C	
5	LBCAT	CHEMISTRY	CHEMISTRY	CHEMISTRY	C	
6	LBCAT	HEMATOLOGY	HEMATOLOGY	HEMATOLOGY	C	
7	LBNRIND	LOW	LOW	LOW	C	
8	LBNRIND	NORMAL	NORMAL	NORMAL	C	
9	LBNRIND	HIGH	HIGH	HIGH	C	
10	NY	NO	NO	N	C	
11	NY	YES	YES	Y	C	
12	VISIT	Baseline	Baseline	Baseline	C	
13	VISIT	3 Months	3 Months	3 Months	C	
14	VISIT	6 Months	6 Months	6 Months	C	
15	XPTEST	Pain Score	Pain Score	Pain Score	C	
16	XPTESTCD	Pain Score	Pain Score	XPPAIN	C	

Reference and Source Data Set Metadata

You need metadata data sets that define the reference standard in the XYZ-SDTM-3.2 core standard, as well as metadata that defines the source variables for your study. The reference standards data sets are already out there from our work in Chapter 4 and can be found at
C:\cstGlobalLibrary\standards\XYZ_program_cdisc-sdtm-3.2-1.6\metadataas
reference_columns.sas7bdat and reference_tables.sas7bdat. However, you need to create the source_columns.sas7bdat and source_tables.sas7bdat data sets for the particular study. These data sets are structured the same as the reference data set versions. We already created versions of these data sets in Chapter 4 when we used the CST to create define.xml. You can just get those data sets from
C:\Users\yourid\Desktop\SAS_BOOK_cdisc_2nd_edition\CST\CST_make_define\metadata and place them under
C:\Users\yourid\Desktop\SAS_BOOK_cdisc_2nd_edition\CST\validate_sdtm\metadata.

Properties Files

You need a copy of the initialize.properties and validation.properties files. You can just copy these from the XYZ123 standard from C:\cstGlobalLibrary\standards\XYZ_program_cdisc-sdtm-3.2-1.6\programs to C:\Users\yourid\Desktop\SAS_BOOK_cdisc_2nd_edition\CST\validate_sdtm\programs.

SDTM Data Sets

The CST is validating your SDTM data sets, so it needs to see them as well. You can place those data sets at C:\Users\yourid\Desktop\SAS_BOOK_cdisc_2nd_edition\CST\validate_sdtm\data.

SAS Clinical Standards Toolkit SDTM Validation Program

Now that the metadata, data, and other files are in place to run the SAS Clinical Standards Toolkit SDTM validation job, you can set that program up. You can copy a template program from C:\cstSampleLibrary\cdisc-sdtm-3.2-1.6\sascstdemodata\programs to your validation area at C:\Users\yourid\Desktop\SAS_BOOK_cdisc_2nd_edition\CST\validate_sdtm\programs. The primary revisions to that program are with the work.sasreferences data set. Here is the revised program:

```
**********************************************************************;
* validate_data.sas                                               *;
*                                                                 *;
* Sample driver program to perform a primary Toolkit action, in   *;
* this case, validation, to assess compliance of some source      *;
* data and metadata with a registered standard. A call to a       *;
* standard-specific validation macro is required later in this    *;
* code.                                                           *;
*                                                                 *;
* Assumptions:                                                    *;
*   The SASReferences file must exist, and must be identified in  *;
*   the call to cstutil_processsetup if it is not                 *;
*   work.sasreferences.                                           *;
*                                                                 *;
* CSTversion  1.6                                                 *;
**********************************************************************;

%let _cstStandard=CDISC-SDTM;
%let _cstStandardVersion=XYZ-SDTM-3.2; *<--- 3.1.1, 3.1.2, 3.1.3, or 3.2 *;
%let _cstCTPath=C:\cstSampleLibrary\cdisc-ct-1.0.0-
1.6\data\sdtm\201312\formats;
%let _cstCTMemname=cterms.sas7bcat;
%let _cstCTDescription=201312;

%cst_setStandardProperties(_cstStandard=CST-
FRAMEWORK,_cstSubType=initialize);

* Set Controlled Terminology version for this process  *;
%cst_getstandardsubtypes(_cstStandard=CDISC-
TERMINOLOGY,_cstOutputDS=work._cstStdSubTypes);
data _null_;
  set work._cstStdSubTypes (where=(standardversion="&_cstStandard" and
                             isstandarddefault='Y'));
  * User can override CT version of interest by specifying a different
where
     clause:          *;
  * Example: (where=(standardversion="&_cstStandard" and
                standardsubtypeversion='201104'))   *;
  call symputx('_cstCTPath',path);
  call symputx('_cstCTMemname',memname);
```

```
    call symputx('_cstCTDescription',description);
run;

proc datasets lib=work nolist;
  delete _cstStdSubTypes;
quit;


***********************************************************************;
* The following data step sets (at a minimum) the studyrootpath and
* studyoutputpath. These are used to make the driver programs portable
* across platforms and allow the code to be run with minimal modification.
* These macro variables by default point to locations within the
* cstSampleLibrary, set during install but modifiable thereafter.  The
* cstSampleLibrary is assumed to allow write operations by this driver
* module.
*;
***********************************************************************;

%cstutil_setcstsroot;
data _null_;
  call symput
('studyRootPath',"C:\Users\yourid\Desktop\SAS_BOOK_cdisc_2nd_edition\CST\va
lidate_sdtm");
  call symput
('studyOutputPath',"C:\Users\yourid\Desktop\SAS_BOOK_cdisc_2nd_edition\CST\
validate_sdtm");
run;
%let workPath=%sysfunc(pathname(work));

%let _cstSetupSrc=SASREFERENCES;

***********************************************************************;
* One strategy to defining the required library and file metadata for a
* CST process is to optionally build SASReferences in the WORK library.
* An example of how to do this follows.
* The call to cstutil_processsetup below tells CST how SASReferences will
* be provided and referenced.  If SASReferences is built in work, the call
* to cstutil_processsetup may, assuming all defaults, be as simple as
* %cstutil_processsetup()
***********************************************************************;

***********************************************************************;
* Build the SASReferences data set
* column order: standard, standardversion, type, subtype, sasref, reftype,
*               iotype, filetype, allowoverwrite, relpathprefix, path,
*               order, memname, comment
* note that &_cstGRoot points to the Global Library root directory
* path and memname are not required for Global Library references -
* defaults will be used
***********************************************************************;
%cst_createdsfromtemplate(_cstStandard=CST-FRAMEWORK,
    _cstType=control,_cstSubType=reference,
_cstOutputDS=work.sasreferences);
proc sql;
  insert into work.sasreferences
  values ("CST-FRAMEWORK"     "1.2"                    "messages"
""                "messages" "libref" "input" "dataset" "N"   ""
"C:\cstGlobalLibrary\standards\XYZ_program_cdisc-sdtm-3.2-1.6\messages" 1
"messages.sas7bdat" "")
  values ("CST-FRAMEWORK"     "1.2"                    "template"
""                "csttmplt" "libref" "input" "folder"  "N"   "" ""
2 ""                 "")
  values ("&_cstStandard"     "&_cstStandardVersion"  "autocall"
""                "sdtmauto" "fileref" "input" "folder"  "N"   ""
```

```
"C:\cstGlobalLibrary\standards\XYZ_program_cdisc-sdtm-3.2-1.6\macros"     1
""              "")
  values ("&_cstStandard"        "&_cstStandardVersion"  "control"
"reference"           "cntl_s"   "libref"  "both"    "dataset"  "Y"   ""
"&workpath"                                   .  "sasreferences.sas7bdat"
"")
  values ("&_cstStandard"        "&_cstStandardVersion"  "control"
"validation"          "cntl_v"   "libref"  "input"  "dataset"  "N"   ""
"&studyRootPath\control"                      .
"validation_control.sas7bdat"  "")
  values ("&_cstStandard"        "&_cstStandardVersion"  "fmtsearch"    ""
"srcfmt"   "libref"  "input"   "catalog"  "N"   ""
"&studyRootPath\terminology\formats"                  1 "formats.sas7bcat"
"")
  values ("&_cstStandard"        "&_cstStandardVersion"  "messages"     ""
"sdtmmsg"  "libref"  "input"   "dataset"  "N"   "" ""
2 ""             "")
  values ("&_cstStandard"        "&_cstStandardVersion"  "lookup"       ""
"lookup"   "libref"  "input"   "dataset"  "N"   "" ""
.  ""             "")
  values ("&_cstStandard"        "&_cstStandardVersion"  "template"     ""
"sdtmtmpl" "libref"  "input"   "folder"   "N"   "" ""
1 ""             "")
  values ("&_cstStandard"        "&_cstStandardVersion"  "properties"
"initialize"          "inprop"   "fileref" "input"  "file"     "N"   ""
"&studyRootPath\programs"                     1 "initialize.properties"
"")
  values ("&_cstStandard"        "&_cstStandardVersion"  "properties"
"validation"          "valprop"  "fileref" "input"  "file"     "N"   ""
"&studyRootPath\programs"                     2 "validation.properties"
"")
  values ("&_cstStandard"        "&_cstStandardVersion"  "referencecontrol"
"validation"          "refcntl"  "libref"  "input"  "dataset"  "N"   ""
"C:\cstGlobalLibrary\standards\XYZ_program_cdisc-sdtm-3.2-
1.6\validation\control" . "validation_master.sas7bdat" "")
  values ("&_cstStandard"        "&_cstStandardVersion"  "referencecontrol"
"standardref"         "refcntl"  "libref"  "input"  "dataset"  "N"   ""
"C:\cstGlobalLibrary\standards\XYZ_program_cdisc-sdtm-3.2-
1.6\validation\control" . "validation_stdref.sas7bdat" "")
  values ("&_cstStandard"        "&_cstStandardVersion"  "referencecterm"
""           "ctref"   "libref"  "input"  "dataset"  "N"   ""
"&studyRootPath\terminology\coding-dictionaries" . "meddra.sas7bdat"
"")
  values ("&_cstStandard"        "&_cstStandardVersion"  "referencemetadata"
"column"              "refmeta"  "libref"  "input"  "dataset"  "N"   ""
"C:\cstGlobalLibrary\standards\XYZ_program_cdisc-sdtm-3.2-1.6\metadata" .
"reference_columns.sas7bdat"  "")
  values ("&_cstStandard"        "&_cstStandardVersion"  "referencemetadata"
"table"               "refmeta"  "libref"  "input"  "dataset"  "N"   ""
"C:\cstGlobalLibrary\standards\XYZ_program_cdisc-sdtm-3.2-1.6\metadata" .
"reference_tables.sas7bdat"   "")
  values ("&_cstStandard"        "&_cstStandardVersion"  "results"
"validationmetrics" "results"  "libref"  "output" "dataset"  "Y"   ""
"&studyOutputPath\results"                    .
"validation_metrics.sas7bdat" "")
  values ("&_cstStandard"        "&_cstStandardVersion"  "results"
"validationresults" "results"  "libref"  "output" "dataset"  "Y"   ""
"&studyOutputPath\results"                    .
"validation_results.sas7bdat" "")
  values ("&_cstStandard"        "&_cstStandardVersion"  "sourcedata"    ""
"srcdata" "libref"  "input"   "folder"   "N"   "" "&studyRootPath\data"
.  ""             "")
  values ("&_cstStandard"        "&_cstStandardVersion"  "sourcemetadata"
"column"              "srcmeta"  "libref"  "input"  "dataset"  "N"   ""
```

```
 "&studyRootPath\metadata"                          . "source_columns.sas7bdat"
 "")
   values ("&_cstStandard"      "&_cstStandardVersion"  "sourcemetadata"
 "table"            "srcmeta"  "libref"  "input"  "dataset"  "N"   ""
 "&studyRootPath\metadata"                          . "source_tables.sas7bdat"
 "")
   values ("CDISC-TERMINOLOGY"  "NCI_THESAURUS"          "fmtsearch"
 ""                   "cstfmt"   "libref"  "input"  "catalog"  "N"   ""
 "&_cstCTPath"                                      2  "&_cstCTMemname"
 "")
   ;
 quit;


 **************************************************************;
 * Debugging aid:   set _cstDebug=1                         *;
 * Note value may be reset in call to cstutil_processsetup  *;
 *  based on property settings.  It can be reset at any     *;
 *  point in the process.                                   *;
 **************************************************************;
 %let _cstDebug=1;
 data _null_;
   _cstDebug = input(symget('_cstDebug'),8.);
   if _cstDebug then
     call execute("options &_cstDebugOptions;");
   else
     call execute(("%sysfunc(tranwrd(options %cmpres(&_cstDebugOptions),
 %str( ),
         %str( no)));"));
 run;


 ********************************************************************************;
 * Clinical Standards Toolkit utilizes autocall macro libraries to contain
 * and reference standard-specific code libraries. Once the autocall path is
 * set and one or more macros have been used within any given autocall
 * library, deallocation or reallocation of the autocall fileref cannot
 * occur unless the autocall path is first reset to exclude the specific
 * fileref. This becomes a problem only with repeated calls to
 * %cstutil_processsetup() or %cstutil_allocatesasreferences within the same
 * sas session. Doing so, without submitting code similar to the code below
 * may produce SAS errors such as:
 *     ERROR - At least one file associated with fileref SDTMAUTO is still
 *             in use.
 *     ERROR - Error in the FILENAME statement.
 *
 * Use of the following code is NOT needed to run this driver module
 * initially.
 ********************************************************************************;
```

```
%*let _cstReallocateSASRefs=1;
%*include "&_cstGRoot/standards/cst-framework-
&_cstVersion/programs/resetautocallpath.sas";

***********************************************************************;
* The following macro (cstutil_processsetup) utilizes the following
* parameters
* _cstSASReferencesSource - Setup should be based upon what initial source?
*    Values: SASREFERENCES (default) or RESULTS data set. If RESULTS:
*       (1) no other parameters are required and setup responsibility is
*           passed to the cstutil_reportsetup macro
*       (2) the results data set name must be passed to cstutil_reportsetup as
*           libref.memname
*
* _cstSASReferencesLocation-The path (folder location) of the sasreferences
*                           dataset (default is the path to the WORK
*                           library)
* _cstSASReferencesName - The name of the sasreferences data set
*                         (default is sasreferences)
***********************************************************************;

%cstutil_processsetup();


***********************************************************************;
* Run the standard-specific validation macro.
***********************************************************************;

%sdtm_validate;

* Delete sasreferences if created above and not needed for additional
processing *;
proc datasets lib=work nolist;
  delete sasreferences / memtype=data;
quit;
```

Note that the work.sasreferences data set refers to a number of the core standard XYZ-SDTM-3.2 standard metadata files as well as some metadata files that we defined in the prior section. When the program has been run, it produces two results data sets found under C:\Users\yourid\Desktop\SAS_BOOK_cdisc_2nd_edition\CST\validate_sdtm\results calledvalidation_metrics.sas7bdat and validation_results.sas7bdat. The validation_metrics data set holds operational statistics on the validation checks. For this run, here is what the first 15 records look like:

	metricparameter	reccount	resultid	srcdata	resultseq
1	# of data sets tested	12	SDTM0001	_ALL_	1
2	# of observations	40	SDTM0001	SRCDATA.AE	1
3	# of observations	60	SDTM0001	SRCDATA.DM	1
4	# of observations	84	SDTM0001	SRCDATA.EX	1
5	# of observations	336	SDTM0001	SRCDATA.LB	1
6	# of observations	66	SDTM0001	SRCDATA.SUPPDM	1
7	# of observations	4	SDTM0001	SRCDATA.TA	1
8	# of observations	1	SDTM0001	SRCDATA.TD	1
9	# of observations	3	SDTM0001	SRCDATA.TE	1
10	# of observations	9	SDTM0001	SRCDATA.TI	1
11	# of observations	42	SDTM0001	SRCDATA.TS	1
12	# of observations	8	SDTM0001	SRCDATA.TV	1
13	# of observations	177	SDTM0001	SRCDATA.XP	1
14	Elapsed time to run check: 0:00:02	.	SDTM0001	CSTCHECK_ZEROOBS	1
15	# of records tested	158	SDTM0011	WORK._CSTSRCCOLUMNMETADATA	1

Here is a snapshot of a few records from the validation_results data set:

	resultid	checkid	resultseq	seqno	srcdata	message	resultseverity	resultflag	_cst_rc	actual
19	CST0100	SDTM0001	1	1	SRCDATA.AE			0	0	
20	CST0100	SDTM0001	1	2	SRCDATA.DM			0	0	
21	CST0100	SDTM0001	1	3	SRCDATA.EX			0	0	
22	CST0100	SDTM0001	1	4	SRCDATA.LB			0	0	
23	CST0100	SDTM0001	1	5	SRCDATA.SUPPDM			0	0	
24	CST0100	SDTM0001	1	6	SRCDATA.TA			0	0	
25	CST0100	SDTM0001	1	7	SRCDATA.TD			0	0	
26	CST0100	SDTM0001	1	8	SRCDATA.TE			0	0	
27	CST0100	SDTM0001	1	9	SRCDATA.TI			0	0	
28	CST0100	SDTM0001	1	10	SRCDATA.TS			0	0	
29	CST0100	SDTM0001	1	11	SRCDATA.TV			0	0	
30	CST0100	SDTM0001	1	12	SRCDATA.XP			0	0	

A resultflag of 0 means that the check ran without problems. A resultflag of 1 means there was a finding.

At this point, the results of the SAS Clinical Standards Toolkit SDTM validation run are in unformatted raw SAS data sets. You can use the CST template program cst_report.sas found in the CST Sample library at C:\cstSampleLibrary\cdisc-sdtm-3.2-1.6\sascstdemodata\programs to render those results in a nicely formatted PDF file. If you run this program right after the validate_data.sas program, all of the necessary references will already be in place. Here is what the cst_report.sas program looks like:

```
**********************************************************************;
* cst_report.sas                                                    *;
*                                                                   *;
* Sample driver program to perform a primary Toolkit action, in     *;
* this case, reporting of process results.  This code performs any   *;
* needed set-up and data management tasks, followed by one or more   *;
* calls to the %cstutil_createreport() macro to generate report      *;
* output.                                                           *;
*                                                                   *;
* Two options for invoking this routine are addressed in these       *;
* scenarios:                                                         *;
*   (1) This code is run as a natural continuation of a CST process, *;
*        within the same SAS session, with all required files        *;
*        available. The working assumption is that the SASReferences  *;
*        data set (referenced by the _cstSASRefs macro) exists and    *;
*        contains information on all input files required for reporting.*;
```

```
*     (2) This code is being run in another SAS session with no CST setup   *;
*         established, but the user has a CST results data set and           *;
*         therefore can derive the location of the SASReferences file that   *;
*         can provide the full CST setup needed to run the reports.          *;
*                                                                            *;
* Assumptions:                                                               *;
*   To generate all panels for both types of reports, the following          *;
*   metadata is expected:                                                    *;
*     - the SASReferences file must exist, and must be identified in the      *;
*       call to cstutil_processsetup if it is not work.sasreferences.        *;
*     - a Results data set                                                   *;
*     - a (validation-specific) Metrics data set                            *;
*     - the (validation-specific) run-time Control data set itemizing        *;
*       the validation checks requested.                                     *;
*     - access to the (validation-specific) check messages data set          *;
*                                                                            *;
* CSTversion  1.3                                                            *;
*****************************************************************************;

%let _cstStandard=CDISC-SDTM;
%let _cstStandardVersion=XYZ-SDTM-3.2;   * <------- 3.1.1, 3.1.2, 3.1.3, or
3.2   *;

%cst_setStandardProperties(_cstStandard=CST-
FRAMEWORK,_cstSubType=initialize);

*****************************************************************************;
* The following data step sets (at a minimum) the studyrootpath and         *;
* studyoutputpath.  These are used to make the driver programs portable      *;
* across platforms and allow the code to be run with minimal                 *;
* modification. These macro variables by default point to locations          *;
* within the cstSampleLibrary, set during install but modifiable              *;
* thereafter.  The cstSampleLibrary is assumed to allow write operations     *;
*   by this driver module.                                                   *;
*****************************************************************************;

%cstutil_setcstsroot;
data _null_;
  call symput
('studyRootPath',"C:\Users\yourid\Desktop\SAS_BOOK_cdisc_2nd_edition\CST\va
lidate_sdtm");
  call symput
('studyOutputPath',"C:\Users\yourid\Desktop\SAS_BOOK_cdisc_2nd_edition\CST\
validate_sdtm");
run;

%let workPath=%sysfunc(pathname(work));

* Initialize macro variables used for this task   *;

%let _cstRptControl=;
%let _cstRptLib=;
%let _cstRptMetricsDS=;
%let _cstRptOutputFile=&studyOutputPath/results/cstreport.pdf;
%let _cstRptResultsDS=;
%let _cstSetupSrc=SASREFERENCES;

***********************************************************;
* Debugging aid:  set _cstDebug=1                         *;
* Note value may be reset in call to cstutil_processsetup *;
*  based on property settings.  It can be reset at any     *;
*  point in the process.                                  *;
***********************************************************;
%let _cstDebug=0;
```

```
data _null_;
  _cstDebug = input(symget('_cstDebug'),8.);
  if _cstDebug then
    call execute("options &_cstDebugOptions;");
  else
    call execute(("%sysfunc(tranwrd(options %cmpres(&_cstDebugOptions),
%str( ), %str( no)));"));
run;

* The call to cstutil_processsetup below tells CST how SASReferences    *;
* will be provided and referenced.  If SASReferences is built in work,  *;
* the call to cstutil_processsetup may, assuming all defaults, be as     *;
* simple as %cstutil_processsetup()                                      *;
*************************************************************************;


*************************************************************************;
* Clinical Standards Toolkit utilizes autocall macro libraries to       *;
* contain and reference standard-specific code libraries.  Once the     *;
* autocall path is set and one or more macros have been used within any  *;
* given autocall library, deallocation or reallocation of the autocall  *;
* fileref cannot occur unless the autocall path is first reset to       *;
* exclude the specific fileref.                                         *;
*                                                                       *;
* This becomes a problem only with repeated calls to                    *;
* %cstutil_processsetup() or %cstutil_allocatesasreferences within the  *;
* same sas session. Doing so, without submitting code similar to the    *;
* code below may produce SAS errors such as:                            *;
* ERROR - At least one file associated with fileref SDTMAUTO is still   *;
*         in use.                                                       *;
* ERROR - Error in the FILENAME statement.                             *;
*                                                                       *;
* If you call %cstutil_processsetup() or %cstutil_allocatesasreferences *;
* more than once within the same sas session, typically using %let      *;
* _cstReallocateSASRefs = 1 to tell CST to attempt reallocation, use of *;
* the following code is recommended between each code submission.        *;
*                                                                       *;
* Use of the following code is NOT needed to run this driver module     *;
* initially.                                                            *;
*************************************************************************;


%*let _cstReallocateSASRefs=1;
%*include "&_cstGRoot/standards/cst-framework-
&_cstVersion/programs/resetautocallpath.sas";


*************************************************************************;
* The following macro (cstutil_processsetup) utilizes the following     *;
* parameters:                                                           *;
*                                                                       *;
* _cstSASReferencesSource - Setup should be based upon what initial     *;
*                           source?                                     *;
*   Values: SASREFERENCES (default) or RESULTS data set. If RESULTS:    *;
*      (1) no other parameters are required and setup responsibility is *;
*          passed to the cstutil_reportsetup macro                      *;
*      (2) the results data set name must be passed to                  *;
*          cstutil_reportsetup as libref.memname                        *;
*                                                                       *;
* _cstSASReferencesLocation - The path (folder location) of the         *;
*                             sasreferences data set (default is the    *;
*                             path to the WORK library)                 *;
*                                                                       *;
* _cstSASReferencesName - The name of the sasreferences data set        *;
*                         (default is sasreferences)                    *;
*************************************************************************;
```

```
%cstutil_processsetup(_cstSASReferencesLocation=&studyrootpath/control);

%*let _cstRptResultsDS=results.validation_results;
%cstutil_reportsetup(_cstRptType=Results);

*************************************************************************;
* Run reports                                                          *;
* Note multiple invocations require unique &_cstreportoutput parameter *;
* values                                                               *;
*************************************************************************;

%cstutil_createreport(
    _cstsasreferencesdset=&_cstSASRefs,
    _cstresultsdset=&_cstRptResultsDS,
    _cstmetricsdset=&_cstRptMetricsDS,
    _cstreportbytable=N,
    _cstreporterrorsonly=Y,
    _cstreportobs=50,
    _cstreportoutput=%nrbquote(&_cstRptOutputFile),
    _cstsummaryReport=Y,
    _cstioReport=Y,
    _cstmetricsReport=Y,
    _cstgeneralResultsReport=Y,
    _cstcheckIdResultsReport=Y);
```

This program creates a file called cstreport.pdf, and the first page looks like this with all of the CST validation checks following it:

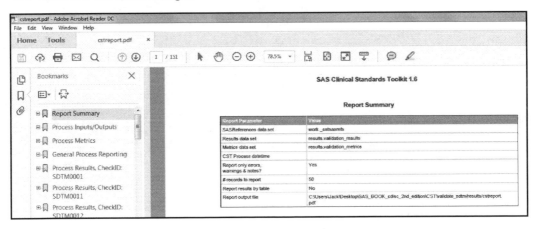

You can also take the validation_results SAS data set and format it however you want via PROC REPORT or any other reporting mechanism that you want to use. There is also the option of defining your own validation checks within the SAS Clinical Standards Toolkit. For information about how to do that, you should refer to the *SAS Clinical Standards Toolkit: User's Guide*.

A similar process may be followed for ADaM data set validation because the Clinical Standards Toolkit supports ADaM as well. However, the results metadata extension is not supported for validation purposes in the toolkit.

SAS Clinical Data Integration SDTM Validation

In Chapter 5, we implemented the SDTM using SAS Clinical Data Integration. Also in that chapter, we imported the SAS Clinical Standards Toolkit SDTM standard for XYZ123, produced the SDTM data sets, and produced a define file. Here we will use SAS Clinical Data Integration to validate the SDTM that we created in Chapter 5. Fortunately, SAS Clinical Data Integration gives you a transformation called **CDISC-SDTM Compliance** that can be dragged to a job. This transformation is easier to implement than the manual setup process found above in the SAS Clinical Standards Toolkit. The transformation looks like this in CDI:

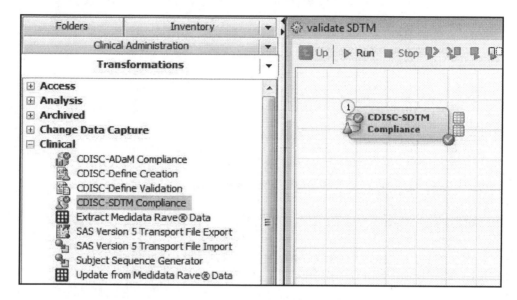

Assuming that the metadata was imported and set up properly in Chapter 5 in SAS Clinical Data Integration, there are only a few additional steps to finish this validation job. If you right-click **CDISC-SDTM Compliance**, select **Properties**, and click the **Data Standard** tab, then you can select the standard to validate your SDTM data against. In this case, we want to select our customized XYZ-SDTM-3.2 standard:

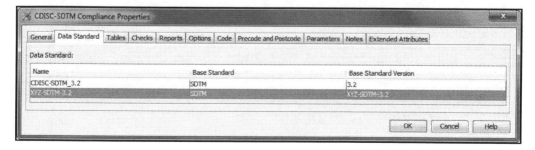

Under the **Domains** tab, we can subset for individual domains if we want, like this:

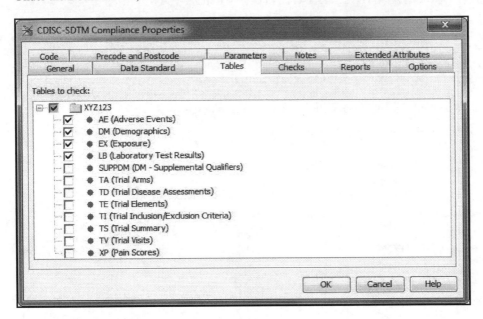

The **Checks** tab enables you to add any of the built-in SDTM checks from SAS, WebSDM, or OpenCDISC that you would like. That screen looks like this:

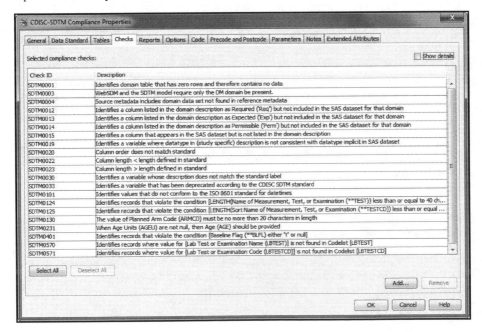

Under the **Reports** tab, you need to specify where you want the SDTM validation reports saved. You can specify that you want the validation reports given to you by domain or by validation check. That screen looks like this:

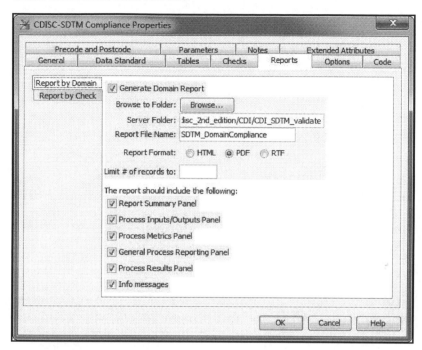

Now you can run the job and review your SDTM validation results in the server folder specified. At the start of the compliance report, we get a nice by-domain bookmarked summarization of findings like this:

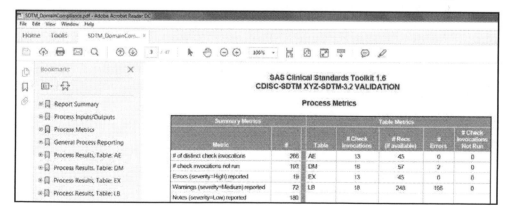

Then if we click the AE domain bookmark, we would see this:

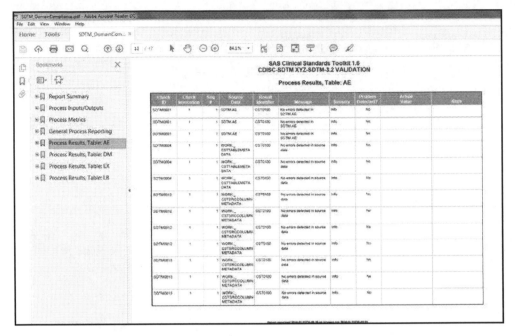

In this example, SAS Clinical Data Integration is serving as a front-end graphical user interface to the SAS Clinical Standards Toolkit.

Because of the volume of checks available and because many of them are informational, it would be wise for you to subset the checks for what your organization deems to be important. The validation reports that SAS Clinical Data Integration can generate via the SAS Clinical Standards Toolkit can be prohibitively large otherwise.

SAS Clinical Standards Toolkit Define.xml Validation Program

In Chapter 4, we used the Clinical Standards Toolkit to create define.xml for our XYZ123 SDTM data. That same define file also needs validation. The SAS Clinical Standards Toolkit 1.6 provides validation of the define file against the published CDISC XML schema to ensure that the define file is well formed XML. In our create_definexml.sas program in Chapter 4, we included this macro call:

```
%cstutilxmlvalidate();
```

If there are any warnings or errors messages of note in the creation of the define.xml file, they would get stored in the C:\Users\yourid\Desktop\SAS_BOOK_cdisc_2nd_edition\CST\CST_make_define\results\write_r esults.sas7bdat data set. SAS Clinical Standards Toolkit 1.7 does go a bit further in providing an additional driver program that will also check the data set and variable-level metadata against the contents of your SDTM data.

SAS Clinical Data Integration Define.xml Validation

In Chapter 5, we used SAS Clinical Data Integration to build our SDTM data sets as well as to generate our define.xml file. Once again, CDI provides us with a nice graphical user interface transformation to enable us to validate the define file that we created. If we look at the clinical transformations available to us, we see **CDISC-Define Validation** as an option as seen here:

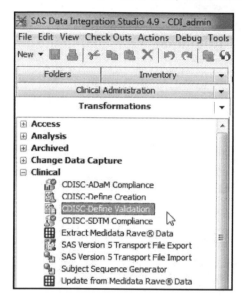

We can add that to the end of our "make define.xml" job with a simple drag:

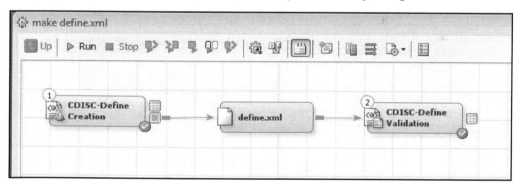

If you open the **Properties** window of the **CDISC-Define Validation** transformation, you can select the version (for example, 1 or 2) of define.xml that you wish, and you can specify the message level (for example, info, warning, error) of the validation reporting. Finally, you need to click the **Reports** tab to specify where to store the results of the define file validation run as follows:

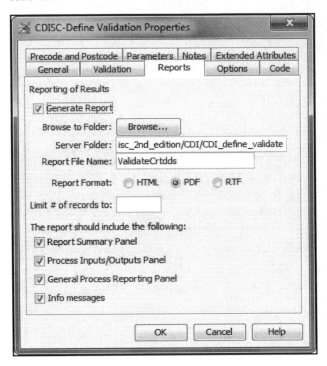

The results of running this transformation are stored in a file called ValidateCrtdds.pdf and look like this:

SAS Clinical Standards Toolkit 1.6
CDISC-DEFINE-XML 2.0.0 XMLVALIDATE CDISC-DEFINE-XML

General Process Reporting

Seq #	Source Data	Result Identifier	Severity	Problem Detected?	Message
1	CST_SETPROPERTIES	CST0108	Info	No	The properties were processed from the PATH c:\cstGlobalLibrary/standards/cst-framework-1.6/programs/initialize.properties
1	CST_SETPROPERTIES	CST0108	Info	No	The properties were processed from the PATH c:\cstGlobalLibrary/standards/cdisc-definexml-2.0.0-1.6/programs/initialize.properties
1	CST_CREATEDSFROMTEMPLATE	CST0200	Info	No	The SAS libref csttmplt was allocated to c:\cstGlobalLibrary/standards/cst-framework-1.6/templates to perform the template lookup
2	CST_CREATEDSFROMTEMPLATE	CST0102	Info	No	work.sasreferences was created as requested
1	CST_INSERTSTANDARDSASREFS	CST0200	Info	No	SASReferences data set was successfully validated
2	CSTUTIL_ALLOCATESASREFERENCES	CST0200	Info	No	SASReferences data set was successfully validated
1	CST_SETPROPERTIES	CST0108	Info	No	The properties were processed from the PATH c:\cstGlobalLibrary/standards/cdisc-definexml-2.0.0-1.6/programs/initialize.properties
1	CST_SETPROPERTIES	CST0108	Info	No	The properties were processed from the PATH c:\cstGlobalLibrary/standards/cdisc-definexml-2.0.0-1.6/programs/validation.properties
1	CSTUTILXMLVALIDATE	CST0200	Info	No	PROCESS STANDARD: CDISC-DEFINE-XML
2	CSTUTILXMLVALIDATE	CST0200	Info	No	PROCESS STANDARDVERSION: 2.0.0
3	CSTUTILXMLVALIDATE	CST0200	Info	No	PROCESS DRIVER: None or unspecified
4	CSTUTILXMLVALIDATE	CST0200	Info	No	PROCESS DATE: 2016-01-23T21:00:24
5	CSTUTILXMLVALIDATE	CST0200	Info	No	PROCESS TYPE: XMLVALIDATE CDISC-DEFINE-XML
6	CSTUTILXMLVALIDATE	CST0200	Info	No	PROCESS SASREFERENCES: C:\Users\Jack\AppData\Local\Temp\SAS Temporary Files_TD5792_JACK-LAPTOP_\Prc2\sasreferences.sas7bdat
7	CSTUTILXMLVALIDATE	CST0200	Info	No	PROCESS STUDYROOTPATH: <not used>
8	CSTUTILXMLVALIDATE	CST0200	Info	No	PROCESS GLOBALLIBRARY: c:\cstGlobalLibrary
9	CSTUTILXMLVALIDATE	CST0200	Info	No	PROCESS CSTVERSION: 1.6
1	JAVA CHECK	CST0200	Info	No	No Java issues
1	CSTUTILPROCESSXMLLOG	CST0200	Info	No	XML Log file C:\Users\Jack\AppData\Local\Temp\SAS Temporary Files_TD5792_JACK-LAPTOP_\Prc2\log3535_xmlvalidate.xml does not contain a <TABLE> or <XMLTransformLog> tag
2	CSTUTILXMLVALIDATE	CST0100	Info	No	No errors detected in the XML file
1	CSTUTIL_SAVERESULTS	CST0102	Info	No	_cdiwrk.VALIDATION_RESULTS was created as requested
1	CSTUTIL_SAVERESULTS	CST0102	Info	No	_cdiwrk.VALIDATION_RESULTS was created as requested

Report generated 2016-01-23T21:00:28 on process run 2016-01-23T21:00:24

Chapter Summary

There are now a handful of options available to CDISC implementers intent on ensuring that their data and define files are compliant with the published standards. The ones covered in this chapter include tools that come with the SAS Clinical Data Standards Toolkit and SAS Clinical Data Integration. Because of changing standards, changing validation rules, and a recognition that not all rules apply to all situations, the chosen tool needs to be flexible enough to allow users to either make customizations or to obtain updates. It should also be easy to install and implement, and the validation results must be easy to interpret. In this chapter, we tried to highlight each of these features so that you can decide which tool would be best to use for your particular study, set of studies, or other situation.

Chapter 9: CDISC Validation Using Pinnacle 21 Community

Pinnacle 21 Community, formerly OpenCDISC Community, is a free tool based on Java. Pinnacle 21 Community has evolved from the original OpenCDISC Validator tool that was first released in 2008 via the website https://www.pinnacle21.net/. Originally, the functionality of the OpenCDISC was focused on performing, summarizing, and reporting on the draft compliance checks that were made publicly available by the FDA in 2008 for the purpose of loading CDISC data into the Janus Clinical Trial Repository. (See http://www.fda.gov/downloads/ForIndustry/DataStandards/StudyDataStandards/UCM190628.pdf.) Although today's version of the tool has many new features, it is still primarily used for validating CDISC data to help ensure compliance with the respective CDISC models and regulatory expectations.

From the first release to the current version, the tool has been remarkably successful within the pharmaceutical industry for three primary reasons: It is easy to obtain and install, it is very easy to use once installed, and it is free. Some of these features are intertwined. The fact that it is free makes it very easy to acquire and install, because there is no licensing to procure beforehand. But the installation procedure itself is also quite simple. Once the tool is installed, the intuitive user interface makes generating a validation report very easy. Once generated, the report itself is organized in an intuitive and easy to comprehend spreadsheet. As of Version 2.1.0 of the tool, validation is available for SDTM data, ADaM data, SEND data, and define.xml files.

In this chapter, we will demonstrate how to use the Pinnacle 21 Community tool using the SDTM, ADaM data, and the define.xml files that we created in previous chapters.

There are still SAS elements within this chapter. We will first cover how SAS can be used to run Pinnacle 21 Community. We will also introduce you to a SAS macro, %adamtrace, that is used to ensure that records that trace back to one another are, in fact, consistent with respect to their values.

Getting Started with Pinnacle 21 Community

Pinnacle 21 Community can be downloaded via the Pinnacle 21 website (https://www.pinnacle21.net/downloads). Instructions on how to install the software are available at https://www.pinnacle21.net/projects/installing-opencdisc-community, and some additional installation details are below.

Running Pinnacle 21 Community (Graphical User Interface)

Pinnacle 21 Community is available for both PCs and Macs. On a PC, you can install the software pretty much anywhere you wish (provided you have Write access to the intended location). In the root installation directory, simply double-click the pinnacle21-community.exe file, and the application will start. The "home" screen displays the four basic functions of the tool: the Validator, the define.xml Generator, the Data Converter, and the ClinicalTrials.gov Miner.

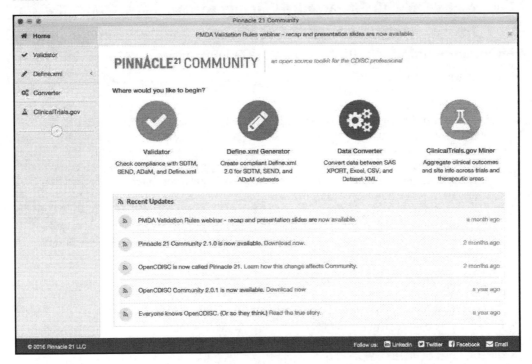

We will focus primarily on the Validator, shown in the next screenshot.

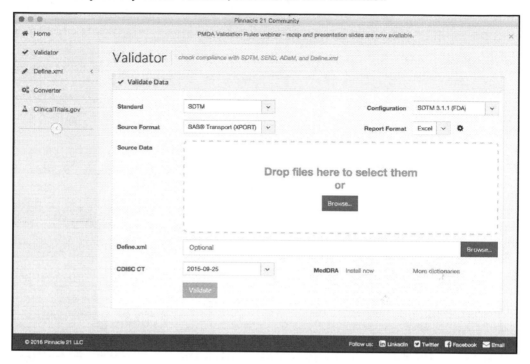

The straightforward interface hardly deserves any explanation, but we'll provide one anyway. First, select the standard that corresonds to what you want to have validated (SDTM, ADaM, SEND, or Define.xml). Next, select the configuration. The available drop-down list depends on the standard chosen. For SDTM, many different versions of the standard are available. Some are designated as FDA checks, and some are designated as PMDA checks (Pharmaceuticals and Medical Device Agency—essentially the Japanese FDA). Typically, SAS transport files will be chosen from the **Source Format** options, but other options are also available. The **Source Data** can either be dragged to the white window or selected manually via the **Browse** button.

For added insurance that your data are consistent with your metadata, there is also the option to select a define.xml file for cross-validation. Note that the findings from a validation that includes a corresponding define.xml file may differ from those that do not include a define.xml file. So it's recommended to do separate validations for both scenarios.

Lastly, the **CDISC CT** pertains to the appropriate version of the CDISC controlled terminology that was used for your data. These versions are stored electronically by the National Cancer Institute (NCI). (See http://www.cancer.gov/research/resources/terminology/cdisc.)

When you are running the Pinnacle 21 Community GUI, the options for the format of the validation report are Microsoft Excel file or a CSV file. Note that you do not necessarily need Excel to open the Excel file. Many common spreadsheet programs can be used to read the Excel file results.

In the following example, the SDTM data that you first created in Chapter 3 are selected for evaluation, and the define.xml file that you created in Chapter 2 is also selected. When all of your inputs have been set, you are ready to start the validation process by clicking the **Validate** button. When the validation is done, you should see a window similar to the following screenshot.

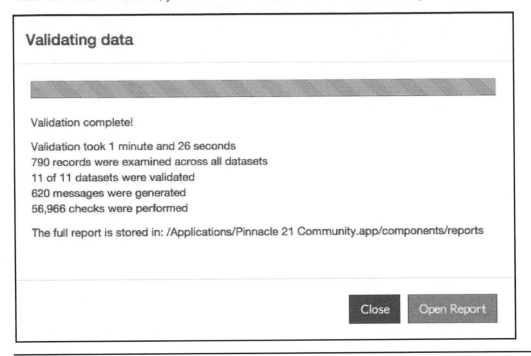

Evaluating the Report

The following screenshot shows a sample of the Excel report early in the validation process. There are four tabs, or worksheets, in the Excel report: **Dataset Summary**, **Issue Summary**, **Details**, and **Rules**. The data set summary provides information about each data set in a quick, snapshot view. Each processed data set is listed in a separate row with columns for the data set label, the SDTM class, filename, and the number of records, warnings, and notices (informational messages) relating to the domain.

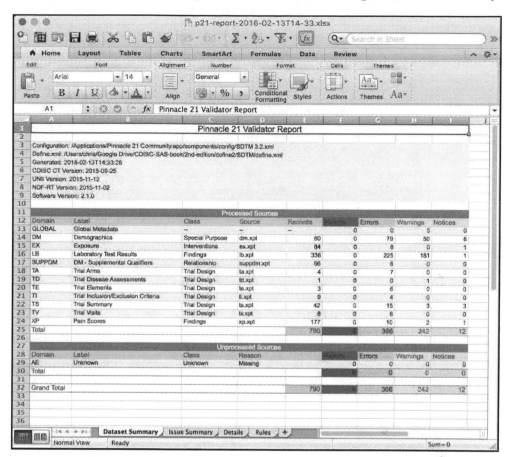

The following screenshot shows a sample of the issue summary. This worksheet lists all errors found and provides a count of each error type. Errors are the most serious issues (unless using PMDA rules, which include a Reject status). A short description of the error is provided as well as a link to the RuleID. Each RuleID is listed in the Rules worksheet, so clicking the link brings you to that worksheet where full details of the rule are provided. These rules are defined with the configuration file that was specified on the start screen, which, in this case, is the SDTM 3.2.xml file. We will cover these configuration files in greater detail in a later section.

	A	B	C	D	E	F
1				Pinnacle 21 Validator Report		
2						
3	Configuration: /Applications/Pinnacle 21 Community.app/components/config/SDTM 3.2.xml					
4	Define.xml: /Users/chris/Google Drive/CDISC-SAS-book/2nd-edition/define2/SDTM/define.xml					
5	Generated: 2016-02-13T14:33:26					
6	CDISC CT Version: 2015-09-25					
7	UNII Version: 2015-11-12					
8	NDF-RT Version: 2015-11-02					
9	Software Version: 2.1.0					
10						
11				Issue Summary		
12	Source	Pinnacle 21 ID	Publisher ID	Message	Severity	Found
13	GLOBAL					
14		SD0061	FDAC024	Domain referenced in define.xml but dataset is missing	Warning	1
15		SD1106	FDAC005	Missing AE dataset	Warning	1
16		SD1108	FDAC007	Missing VS dataset	Warning	1
17		SD1110	FDAC003	Missing DS dataset	Warning	1
18		SD1111	FDAC008	Missing SE dataset	Warning	1
19	DM					
20		CT2001	FDAC340	DTHFL value not found in 'No Yes Response (Yes only)' non-extensible codelist	Error	60
21		SD1082	FDAC036	Variable length is too long for actual data	Error	19
22		SD0006	FDAC113	No baseline result in LB for subject	Warning	50
23		SKIP_SD0006		EG is missing or lacks necessary variables and cannot be used for this cross-dataset validation	Notice	1
24		SKIP_SD0006		MB is missing or lacks necessary variables and cannot be used for this cross-dataset validation	Notice	1
25		SKIP_SD0006		MS is missing or lacks necessary variables and cannot be used for this cross-dataset validation	Notice	1
26		SKIP_SD0006		PC is missing or lacks necessary variables and cannot be used for this cross-dataset validation	Notice	1
27		SKIP_SD0006		VS is missing or lacks necessary variables and cannot be used for this cross-dataset validation	Notice	1
28		SKIP_SD0069		DS is missing or lacks necessary variables and cannot be used for this cross-dataset validation	Notice	1
29	EX					
30		SD1082	FDAC036	Variable length is too long for actual data	Error	8
31		SKIP_SD0082		DS is missing or lacks necessary variables and cannot be used for this cross-dataset validation	Notice	1
32	LB					
33		CT2003	FDAC342	LBTESTCD and LBTEST values do not have the same Code in CDISC CT	Error	180
34		SD0037	FDAC057	Value for LBTEST not found in (LBTEST) user-defined codelist	Error	30
35		SD1082	FDAC036	Variable length is too long for actual data	Error	15
36		CT2002	FDAC341	LBTEST value not found in 'Laboratory Test Name' extensible codelist	Warning	180
37		SD1076	FDAC031	Model permissible variable added into standard domain	Warning	1
38		SKIP_SD0065		SV is missing or lacks necessary variables and cannot be used for this cross-dataset validation	Notice	1

Dataset Summary | **Issue Summary** | Details | Rules | +

Normal View   Ready   Sum=0

After errors, a list of all warnings is provided (within each data set). Warnings are considered less severe. But they should be evaluated to determine whether they are issues that require some sort of fix either to the data, the metadata, or perhaps both.

After the warnings are listed, notices are provided. These often include the checks that could not be made due to a lack of data.

The next worksheet is titled Details and is shown in the following screenshot. Here you can find the details pertaining to each rejection, error, warning, or notice (as applicable). These results are listed by domain and provide information about the record number; the affected variables; the pertinent values; the rule ID (with a link to the rule); the message; the category of the rule (Format, Limit, Metadata, Presence, System, or Terminology); and the type of result (Rejection, Error, Warning, or Notice). If you chose to have your report produced in CSV format, only the information seen in the Details worksheet of the Excel report is provided in the resulting CSV report file.

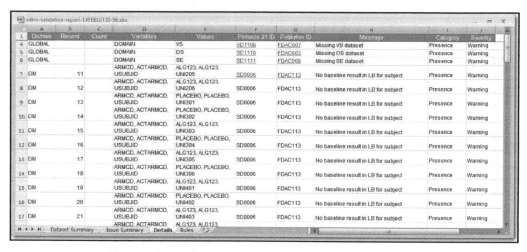

The final worksheet entitled Rules contains a list of each RuleID. The links to each rule ID in the Issues Summary and Details worksheets take you to this worksheet:

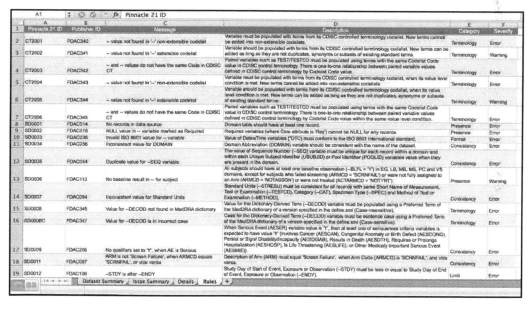

Sometimes certain results, such as warnings (but, in some cases, errors as well) are not pertinent to your particular set of data or for your interpretation or your organization's interpretation of the standard. If this is the case, you can either choose to do one of the following: 1) ignore the result and address it in your reviewer's guide with an explanation as to why the message does not need to be addressed or 2) modify or de-activate the rules. In many cases, the former option would be the preferred one. But in other situations, such as those that might involve a change to the standard that has not yet been reflected in the configuration file, an erroneous check, or a change to rules themselves, the latter option might be preferred. The next section explains how to make changes to the configuration file and related data.

Modifying the Configuration Files

As mentioned earlier, the configuration files that Pinnacle 21 Community uses for performing validation checks are XML files located in the components\config subdirectory of the software's installation directory. With the style sheets already a part of the installation, these files should be viewable in a web browser, as shown in the following screenshot. You can also view the compliance check repository on the Pinnacle 21 website: (https://www.pinnacle21.net/validation-rules).

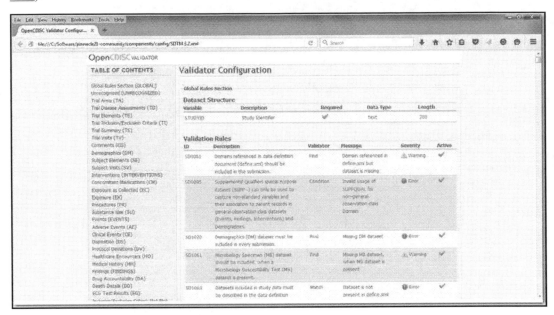

As you review the results of the compliance checks, you might come across checks that, for whatever reason, do not apply to your data, your study, your development program, or you just consider them erroneous. As an alternative to performing these checks and presenting them in your validation report, you could identify them within the XML configuration file and either comment them out or set their active status to **No**. Note that editing of the checks cannot be done via a web browser because web browsers only enable you to view the files. You would first need to open the files in either a text editor or an XML editor. Then you would have to search on checks by the ID and make the appropriate modifications. You should consider saving the validation file under a different name so as not to overwrite the original. If the new file is saved in the same config directory, then it should appear in the drop-down list of configuration files in the GUI.

Deactivating rules by editing the XML configuration file is very straightforward. Let's use one in our validation report as an example. Many of them pertain to check SD1082, "Variable length is too long for actual data." In Chapter 11, we have a macro to address this . But, nonetheless, to call this an actual error might be considered a "stretch" of some valid FDA advice, which in spirit is a preference against gross variable length settings that can inflate data set sizes unnecessarily (like setting the length for an SDTM DOMAIN variable to $200). To deactivate this rule, first simply find it in the configuration file, as shown here:

```
<val:ValidationRuleRef RuleID="SD1082" Active="Yes"/>
```

Unfortunately, this appears in 39 different places within the SDTM 3.2.xml configuration file. But with a text-editor with a global search-and-replace function, you could easily change this:

```
<val:ValidationRuleRef RuleID="SD1082" Active="No"/>
```

Be sure to keep track of the changes made so that they can be communicated to other organizations, including regulatory agencies, who may be using default configuration files for their own internal validation.

Adding new compliance checks is an option, too, although the syntax and logic involved in writing new checks is not a straightforward exercise. An overview of the syntax for the validation rules can be found online at https://www.pinnacle21.net/projects/validator/opencdisc-validation-framework. For example, suppose we want to add one of the SDTM 3.1.2 Amendment 1 variables to the DM domain, RFXSTDTC. This variable represents the date/time of first study treatment exposure. (Note that this and many other Amendment 1 variables were added to SDTM version 3.1.3 and the corresponding Pinnacle 21 configuration files.) We would first need to add this variable to the list of DM variables. So as not to have to renumber all of the existing variables, the variable will be added after the last DM variable in the SDTM 3.1.2 configuration file, DM.RFENDY, as shown in the following code:

```
<ItemRef ItemOID="IT.DM.RFENDY" OrderNumber="26" Mandatory="No"
Role="Timing" val:Core="Model Permissible"/>

<ItemRef ItemOID="DM.RFXSTDTC" OrderNumber="27" Mandatory="No"
Role="Record Qualifier" val:Core="Expected"/>
```

Next, we want to define the label for this variable. Because it is similar to RFSTDTC, we will insert it after that:

```
<ItemDef OID="DM.RFSTDTC" Name="RFSTDTC" DataType="datetime"
def:Label="Subject Reference Start Date/Time"/>

<ItemDef OID="DM.RFXSTDTC" Name="RFXSTDTC" DataType="datetime"
def:Label="Date/Time of First Study Treatment Exposure"/>
```

Finally, we will add a check similar to the check for RFSTDTC. The ID for this check will come sequentially after the last check in the list (SD2265):

```
<val:Required ID="SD0087" Variable="RFSTDTC" When="ARMCD @neqic
'SCRNFAIL' @and ARMCD @neqic 'NOTASSGN'" Message="RFSTDTC cannot be
null for randomized subject" Description="Subject Reference Start
Date/Time (RFSTDTC) is required for all randomized subjects, those
where Planned Arm Code (ARMCD) is not equal to 'SCRNFAIL' or
'NOTASSGN'." Category="Consistency" Type="Warning"/>

<val:Required ID="SD2265" Variable="RFXSTDTC" When="EXSTDTC != ''"
Message="RFXSTDTC cannot be null for subjects who received study
drug" Description="Subject First Study Drug Exposure Date/Time
(RFXSTDTC) is required for all subjects who received protocol-
specified treatment or therapy." Category="Consistency"
Type="Warning"/>
```

If this check is done correctly, you should be able to find it in your validation report.

If the task of writing your own compliance checks looks too daunting, consider communicating with Pinnacle 21 developers via their online web forums. This way you can get additional details about how best to either create your own checks or have the developers consider new checks for general distribution. Another option is to contact CDISC's SDS Compliance Subteam.

One last note about creating your own customized configuration files: You might want to share them with other organizations, including regulatory agencies, who might be doing their own validation. This can be easier than having to repeatedly explain checks from one submission to the next that are not, for whatever reason, pertinent. Fortunately, the Pinnacle 21 validator software is very intelligent in that it will identify in its **Configuration** drop-down menu any file that exists in the proper directory. So providing the customized configuration file along with instructions as to where to put it makes it very easy for other organizations to use your configuration for their own validation. See the following screenshot for our SDTM 3.2-custom file as an example.

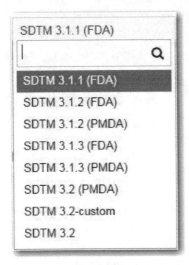

A Note about Controlled Terminology

One common set of warnings that users might experience is that relating to terminology. Staying up-to-date with the most recent controlled terminology is important, but not always practical. When a define.xml file is included when validating SDTM data, the terminology specified in the define file is used to check against values in the data. The configuration files that Pinnacle 21 Community uses for performing validation checks are XML files. The XML files that contain the controlled terminology are located in the components\config\data\CDISC\ subdirectory of the Pinnacle 21 Community installation directory. From there, additional subdirectories exist for SEND, SDTM, and ADaM data. And, within each of those, exists the last branch of subdirectories that are named after the dates that correspond to the version of the controlled terminology that your submission or study data follow. See the following screenshot for an example.

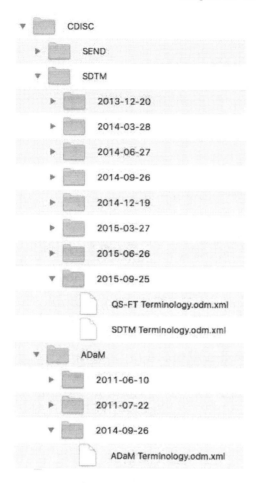

In earlier versions of the software, you could manually add updated XML files to these directories when new controlled terminology versions were added to the National Cancer Institute (NCI) archive at http://evs.nci.nih.gov/ftp1/CDISC/SDTM/ (for SDTM data, for example). Starting with version 2.0 of the software, however, these files are now pushed by Pinnacle 21. This is because there is some pre-processing that has to be done to the XML files in order to make them compatible with the software.

Running Pinnacle 21 Community (Command-Line Mode)

A lesser-known feature of Pinnacle 21 Community is the ability to run it in command-line mode. In command-line mode, a third output option is available for the report file: XML. To the SAS programmer, a possibly better reason to use the tool through the command line is that it provides the ability to run it as a part of a batch process. Before demonstrating this, let's first look at how the tool can be run from the command line in a Windows environment.

To start, open a Command Prompt window. From the window, navigate to the folder where your Pinnacle 21 Community files were installed (or unzipped). Then navigate two folders down to the components\lib directory. A good place to start is to use the -help option. To view this, enter the following command:

```
>java -jar validator-cli-2.1.0.jar -help
```

Note that the filenames are specific to the version of the software being used. So if you are using a newer version of the software, then the filename above will need to be updated accordingly.

The resulting output looks like this:

The -source, -config, and -report parameters are required if you are not using the validator to construct a define.xml file.

Now let's look at each option needed to validate a set of version 3.2 SDTM transport files that are contained in one directory. Table 9.1 provides a summary of these options.

Table 9.1: Selected Pinnacle 21 Community Command-Line Parameters

Option Name	Option Value	Purpose
-type=	SDTM|ADaM|SEND|Define| Custom	Indicates what type of validation is being requested, that of SDTM, ADaM, or SEND data sets or a define.xml file.
-source=	" [dataset -path]*.xpt"	Provides the location of the data sets to be validated. An asterisk (*) can be used as a wildcard to validate all XPT files in the directory.
-config=	" [config-file-path]\config-sdtm-3.2.xml"	Provides the path of the XML configuration file. Unless a customized file is stored elsewhere, this will typically be the location in the components\config subdirectory to where the software was installed or unzipped.
-config:define=	"[define-path]\define.xml"	It is always best to have the data sets validated against the metadata. This option specifies the location of the define.xml file (usually in the same directory as the SDTM data sets).
-report=	" [report-file-path]\sdtm-validation-report-[Date:Time].XLS"	Specifies the location and name of the output Excel file. As a convention, the creation date and time can be appended to the filename.
-report:overwrite=	"yes"	With the date and time appended to the end of the filename, this option might not be needed. Otherwise, with the value of "yes," the old file will be overwritten.

By putting these options together, you can run the Pinnacle 21 Community Validator just as you would through the GUI. However, manually entering all of these commands and file paths is always prone to error. The advantage of using the command-line interface is that, rather than entering the commands yourself each time you want to submit a job, you can enter them in a SAS program and have the SAS program submit the job via the X command. The X command allows operating system commands to be executed during execution of your SAS program. In a Windows environment, this is done via the Command Prompt or DOS window. These operating-system commands can either be run concurrently, while SAS continues to run (when the NOXSYNC option is active), or sequentially, where SAS waits for the Command Prompt window to exit first (when the XSYNC option is active).

The following code is a demonstration of this implementation, specific to an SDTM validation:

```
** Run the Pinnacle 21 Community in batch mode;
** set the appropriate system options;
options symbolgen xsync noxwait ; ❶

** specify the path and the file for running the validator in
command-line mode;
%let p21path=c:\software\pinnacle21-community;
%let validatorjar=components\lib\validator-cli-2.1.0.jar;

** specify the location of the data sets to be validated. ;
%let sourcepath=g:\2nd-edition\define2\sdtm;

** specify the file to be validated;
** to validate all files in a directory use the * wildcard;
%let files=*.xpt;  ❷

** specify the config file path;
%let config=&p21path\components\config\config-sdtm-3.1.2.xml;

** specify the name and location of the codelist file;
%let codelists=&p21path\components\config\data\CDISC\SDTM\2014-09-
26\SDTM Terminology.odm.xml;

** specify the output report path;
%let reportpath=&sources;

** specify the name of the validation report;
** append the file name with the date and time; ❸
%let reportfile=sdtm-validation-report-
&sysdate.T%sysfunc(tranwrd(&systime,:,-)).xls;

** run the report;
x java -jar "&p21path\&validatorjar" -typ=SDTM -
source="&sourcepath\&files " -config="&config"
-config:define="&sources\define.xml" -config:codelists="&codelists"
-report="&reportpath\&reportfile"  -report:overwrite="yes" ; ❹
```

❶ When the XSYNC option is active, SAS waits for the command to finish executing before returning to the SAS commands. With NOXWAIT active, an EXIT command does not have to be entered in order to return control to SAS.

❷ The Pinnacle 21 command line allows the use of a wildcard to evaluate all XPT files in a directory (for example, *.XPT). Alternatively, an individual file can be entered.

❸ An advantage of the command line submission is the ability to specify the name of the report file. We will, however, maintain the GUI convention of appending the date and time to the base filename in order to ensure uniqueness of the filenames.

❹ Finally, all of the options are put together into one X command. Note the use of the config:define option, where it is assumed that the define.xml file exists in the same directory as the data sets themselves. This could easily be changed by adding an additional macro variable for the location of the define file.

For anyone with experience going through the SDTM validation process, the advantages of this approach are rather clear. Rarely is the validation process a simple one- or two-step approach of running a report, addressing the issues, and then running the report again to show that the issues have disappeared. Often this is much more an iterative process where you must whittle away at issues. With this in mind, the advantages of this example are much more apparent. Rather than having to manually validate at each iteration, the validation can be done automatically, as part of the iterative process. Hence, the same rationale for writing and saving SAS programs can be applied to the validation process.

After a version of the code used in this example is customized for your specific project, it could be further generalized for any project. Whatever internal process you might have for setting up new studies and (presumably) specifying data libraries and file paths in a central location can easily be applied to the code.

To help you get started, a SAS macro for automated batch submissions (called %run_p21v) can be found in Appendix D and with the book's online materials under the authors' pages (http://support.sas.com/publishing/authors/index.html). To get the macro to work in your environment, you will likely need to make some updates such as to the software's installation path. Such changes should be minimal, and the macro can be put to use with few modifications. The details of the macro are a bit too involved to include within this chapter, but the example above gives you an idea of the key parameters needed for most common situations. An example macro call is shown below.

```
** validate the data sets;
%run_p21v(type=SDTM, sources=&path\SDTM\, files=*.xpt, define=Y,
ctdatadate=2014-09-26);
```

ADaM Validation with Pinnacle 21 Community

Shortly after the release of version 1.0 of the ADaM IG, a CDISC ADaM validation sub-team was charged with the task of creating a list of ADaM compliance checks. Seeking to avoid the issue with the SDTM compliance checks—having many different sets of checks published by many different organizations—the ADaM team thought it best that checks first be published by the group of individuals who know the standard best—the ADaM team members. As of this writing, Version 1.3 of these checks (released on March 16, 2015) are the most recent and include over 250 checks They are available for download from http://www.cdisc.org/standards/foundational/adam.

While the ADaM team has been working on their own checks, the developers for Pinnacle 21 were also working on a list of additional checks that could be run in their software, similar to the SDTM checks. Efforts have been made to harmonize these two sets of checks into one.

The process by which ADaM validation is done in Pinnacle 21 Community is identical to that by which SDTM checks can be run. You open the tool; you select your standard, source data, and configuration file; and simply click **Validate**.

The next section will address a validation step that is outside the scope of the Pinnacle 21 Community software: ADaM traceability checks.

ADaM Traceability Checks with SAS

As discussed in the previous section, compliance checks of ADaM data follow a process similar to SDTM compliance checks. A component of ADaM validation that is more unique to ADaM data is traceability. With the traceability features built into ADaM, it is possible to do automated checks between the ADaM AVAL in a BDS-structured data set (or AEDECOD in an ADAE file) and a corresponding value in a predecessor data set (for example, an SDTM data set). Doing such checks helps ensure the accuracy of your derivations and of the information that you provide for making those derivations traceable, which can be particularly important for legacy data conversions, as noted in the FDA's Study Data Technical Conformance Guide (http://www.fda.gov/downloads/ForIndustry/DataStandards/StudyDataStandards/UCM384744.pdf).

There are two primary ways that you can reference specific source records in a predecessor data set: 1) by using the SRCDOM, SRCVAR, and SRCSEQ variables or 2) by using an SDTM --SEQ variable. The %adamtrace macro below identifies when these methods have been used and, if they have, checks to see that AVAL in the given ADaM data set matches the source value in the predecessor data set.

If Method 1 has been used (use of SRC--- variables), then the value of the variable that is identified by SRCVAR (in the domain or data set identified by SRCDOM and for the row or record identified by SRCSEQ) is compared to the AVAL for the given data set.

If Method 2 has been used, then the source domain is identified by the prefix of the --SEQ variable present in the data set. The variable used for comparisons is the --STRESN variable. Therefore, if ADLB contains LBSEQ, then the value of AVAL is compared to LBSTRESN for the given subject and the specified LBSEQ value in the LB SDTM domain.

Note, however, that it is assumed that a data set named with a matching prefix exists. If, for example, the lab data were broken down into separate data sets, one for chemistry data named LBC and one for hematology data named LBH, then the macro would not work because of the non-existence of a data set named LB.

If the ADaM data set has ADAE in the name, then both the source variable and the analysis variable in the current data set are both assumed to be AEDECOD, and the values of AEDECOD are compared to one another. Note that the macro has not yet been extended to the new ADaM class OCCDS (Occurrence Data Structure).

The full code to a macro to perform such checks can be found under the authors' pages on the SAS website (http://support.sas.com/publishing/authors/index.html). The macro has only three parameters: ADAMLIB, SRCLIB, and DATASETS. The ADAMLIB parameter should be a libref to a set of ADaM data sets (or, at least, one ADaM data set). It works only on native SAS data sets as opposed to SAS transport files (suggestions on how to easily convert between the two formats can be found in Chapter 11). Similarly, the SRCLIB parameter should refer to a libref for a set of source data sets (typically SDTM data sets). These two parameters are required. The final parameter, DATASETS, is optional. If it is left blank, then every data set in the ADAMLIB is checked to determine whether it contains the traceability variables. If so, the traceability check is performed. Optionally, you can specify one or more data sets (separated by a pipe or | character), and only those specified data sets will be evaluated.

An example macro call to test all applicable data sets and analysis variables (AVAL or AEDECOD) in a given ADaM library with a libref of ADAM and an SDTM source library called SDTM would be as follows:

```
%adamtrace(adamlib=adam, srclib=sdtm);
```

It should be pointed out that the "immediate predecessor" data set, to use the parlance of the ADaM IG, will not always be an SDTM data set. As shown with our example data, where ADTTE was derived from ADAE, the immediate predecessor could be another ADaM data set. If this is the case, then those specific data sets that have SRCDOM pointing to another ADaM data set should be explicitly included in the DATASETS parameter, and the SRCLIB parameter should point to the same library as the ADAMLIB parameter. This is shown in the following example macro call:

```
%adamtrace(adamlib=adam, srclib=adam, datasets=ADTTE);
```

Similarly, you would then have to be careful not to do an additional run of the macro where the DATASETS parameter was left blank and the SRCLIB parameter pointed to the SDTM library. This would cause all ADaM data sets to be checked, including those that reference another ADaM data set via SRCDOM, which would, in turn, cause an error if the referenced ADaM data did not exist in the SDTM library. So for our example, a second macro call would look like this:

```
%adamtrace(adamlib=adam, srclib=sdtm, datasets=ADAE|ADEF);
```

Because ADSL does not have the required traceability variables, it is not listed.

When you use slightly modified versions of the ADaM data created in Chapter 7, results of the %adamtrace macro (from the SAS data set it creates) appear in the following screenshot. An intentional error was added so that there would be a result to view. (However, it should be pointed out that the macro turned out to be quite helpful in identifying unintentional errors when the data sets for this book were first being developed.)

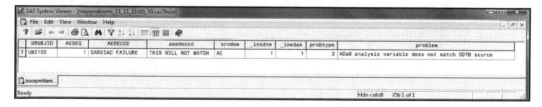

As you can see, results are fairly straightforward. There are columns for the following information:

- The unique subject ID (USUBJID)
- The applicable sequence variable (note that one column is added for each unique sequence variable)
- The analysis variables (AEDECOD and AVAL)
- The source variables
- The source domain (SRCDOM)
- Indicators as to whether a record exists in either the source or the ADaM data

- A code for the type of problem found (PROBTYPE)
- A description of the problem (PROBLEM)

Note that in the case of AE data, the analysis variable is also the source variable. In order to differentiate the two, AEDECOD in the source data (AE) is renamed to AAEDECOD.

There are only two types of problems that are pointed out. The codelist for PROBTYPE and the corresponding values of PROBLEM are as follows:

1 = ADaM record not found in source data

2 = ADaM analysis variable does not match source data

Where the true problem lies tends to involve some detective work. Experience with the macro has shown that adding traceability to the data is not just an added burden. Used in conjunction with an automated tool such as the %adamtrace macro, checking source values can help you uncover problems in your code.

Define.xml Validation with Pinnacle 21 Community

Although some may be content by simply developing a define.xml file that can render in a web browser, the unfortunate news is that getting your file to render is sometimes only half the battle. Ensuring that your file is compliant both with the ODM schema and the Define 2.0 specification is another matter, as is ensuring that your metadata are consistent. How important this type of validation is to a regulatory reviewer is not totally clear, since many of the findings from a Pinnacle 21 Community validation report pertain more to the nuts and bolts of the define file than to the actual data. There is however, always the potential to uncover metadata inconsistencies or missing elements that certainly could contribute to a less efficient regulatory review.

The following screenshot shows the entries for an ADaM define file validation. You will notice many similarities to an SDTM or ADaM data set validation and some differences. The choices for the configuration files are narrowed down to only two options: "Define" and "Define (PMDA)." The number of checks performed between these two options is the same. The primary difference is that some PMDA validation rules result in a "Reject" status rather than just an "Error" or "Warning" in the base Define. There is no need to explicitly indicate whether you are validating a define.xml file for SDTM or ADaM data. This is determined somewhat implicitly by indicating the **CT Standard**, which has two options, SDTM or ADaM, that in turn determine the **CDISC CT** versions that are available.

If all goes well, clicking the **Validate** button will generate a validation report that is very similar to those you have seen with SDTM and ADaM data set validation. If by chance your define file somehow causes the entire software to crash before a report can be generated, consider trying an older version of the software, such as the OpenCDISC Community Validator.

An example Issue Summary from an early version of the validation report is shown in the following screen shot. As you may be able to see from the issues, there are a few that have to do with XML metadata. Many of these are actually related to the fact that the configuration file used for validation does not yet support Analysis Results Metadata. These are good examples where one could either alter existing checks, create new ones, or de-activate some in the configuration file.

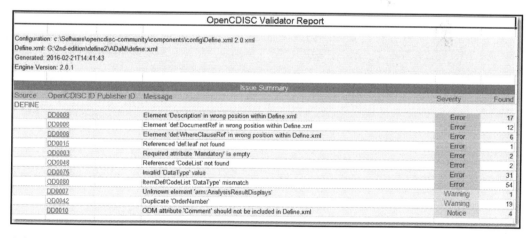

Other checks may be found to be more worthwhile and may even help you to discover inconsistencies between your data sets and metadata that earlier validation exercises did not reveal. Ensuring compliant ODM and a define.xml file that complies with the specifications can also help with your piece of mind prior to a regulatory submission.

Chapter Summary

Pinnacle 21 Community is a free, easy-to-use, open-source software application that provides an alternative to the validation tools provided by SAS. It produces thorough and user-friendly validation reports that can be used to check your SDTM, ADaM, and SEND data as well as your define.xml file. Since it is open-source it is, to some extent, customizable. With its command-line functionality, you can theoretically generate and validate both your data and metadata in one batch process.

A part of the validation process that is unique to ADaM is checking the accuracy of the traceability provided in ADaM data. For this task, the %adamtrace macro was introduced and its use demonstrated. Together with the other validation tools, this piece can be added to your ADaM validation process.

Chapter 10: CDISC Data Review and Analysis

As mentioned in the first chapter of this book, those who stand to gain the most from adapting to CDISC standards are the true users of clinical trial data—those who are most involved with evaluating the safety and efficacy of medical products and who are responsible for assessing the results of a trial. Until now, this book has primarily focused on the process of getting clinical trial data into a CDISC format. In this chapter, however, we will focus on the benefits that you can reap when evaluating data that are already in a CDISC format.

There are three primary benefits that the users of CDISC data stand to gain:

1. Instant familiarity with the structure and format of the data with which they are working
2. Analysis data that is both "one PROC away" and traceable to the source
3. Ability to develop and use software and tools built around the standards

The first benefit is especially true for regulatory reviewers who, until recently, have had to learn entirely new data formats from each new NDA or BLA submission to which they were assigned. The second benefit is specific to ADaM data. Sections of this chapter cover some examples of how ADaM data can be used to quickly produce analysis results.

The third benefit to users of CDISC data was, initially, a hypothetical benefit. From the early stages of development, the greatest promise of data standards was the ability to develop software that could automatically perform many of the routine tasks associated with clinical trial data analysis and review. JMP Clinical is one such tool that has been developed to fulfill this promise. The initial sections of this chapter provide a tutorial to JMP Clinical 5.0 (using JMP 11) and show many examples of the tasks that JMP Clinical can perform.

Safety Evaluations with JMP Clinical

JMP has long been used as a data exploration tool either alongside SAS or completely independent of SAS. It is menu-driven, which makes it easy to learn and use for those who are not statistical programmers. And it has a scripting language, which enables advanced users to write code or scripts that can be reused for repetitive tasks. It has seamless integration with SAS, which obviates the need to convert SAS data sets to a different format, and it enables you to incorporate SAS procedures into a JMP analysis. Lastly, JMP has nice interactive graphics that can be both menu-driven (and, therefore, easy to generate) and customizable.

With these traits, JMP is well suited for leveraging the advantages of CDISC data for the purposes of data exploration. As such, JMP Clinical has been developed as an add-on package to JMP. With JMP Clinical installed, a new menu item, appropriately titled **Clinical**, is added to the JMP menu bar. From this menu item, users have at their disposal a wealth of built-in processes for evaluating patient demographics; producing patient profiles and patient narratives; running routine summaries of adverse events; and using cutting-edge techniques and graphical tools for conducting explorations into a medical product's safety. Most of all, these tasks can be run automatically because they rely on the presence of SDTM data (and ADaM's ADSL) and can therefore harness the knowledge of how the data are organized and structured simply by virtue of the fact that JMP Clinical works only on CDISC data.

Getting Started with JMP Clinical

By default, opening JMP Clinical brings you to the Clinical Starter window. The following screenshot displays this window, along with the options available in the **Clinical** menu item. While the other menu items are standard to JMP, the **Clinical** menu item is unique to JMP Clinical. A common first step when using JMP Clinical is to register a new study. This can be done by clicking the **Add Study from Local Folders** button in the Clinical Starter window, and then entering the study name and information about the location of both the SDTM and ADaM data sets. After this information is entered, JMP will know where to find all of the necessary data sets relating to a study. A logical next step would be to select **Check Required Variables**, shown under **Clinical ▶ Studies**. This is not a CDISC compliance check. Rather, it is primarily a check for the existence of certain variables that JMP Clinical requires for performing automated tasks.

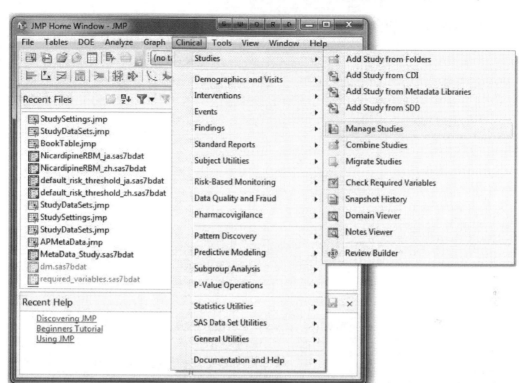

JMP Clinical comes pre-installed with CDISC-compliant study data. These data are from a large clinical trial conducted in the 1980s for a cardiovascular drug called nicardipine, which was subsequently approved to treat hypertension. This pivotal trial randomized more than 900 subjects with subarachnoid hemorrhages, approximately half of whom were randomized to receive nicardipine, while the other half received a placebo. In this chapter, this study is used to demonstrate the tools that JMP Clinical can provide, including the required variable check.

Selecting **Check Required Variables** from the Clinical Starter window creates a new window shown in the following screenshots. Under the **General** tab, you can choose the study for which you wish to run the report and choose the format of the resulting report file (RTF or PDF).

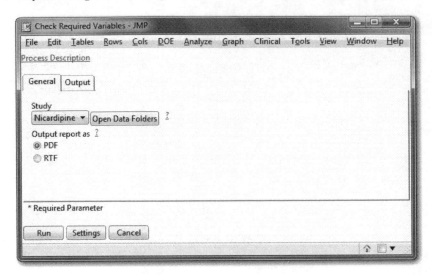

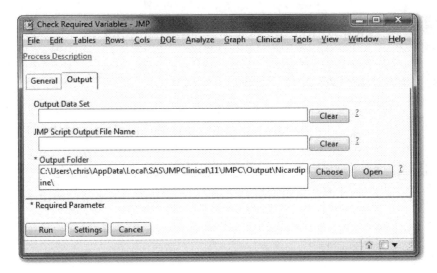

On the **Output** tab, you can specify the names of the resulting data sets and JMP script files and the folder name for all related outputs.

Running the required variable check creates the specified report file (in RTF or PDF format) and a results window in JMP.

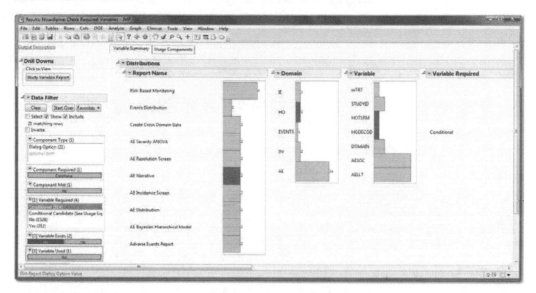

The results window is a dynamic report. Selecting, for example, the AE Narrative report (to be discussed later) and then the 'No' button under "Variable Exists" in the Data filter panel, you can view the missing variables that the report would use if available. In this case, the HO domain is missing altogether, so the variables HOTERM and HODECOD will be available for the narrative reporting. Clicking around this report can provide additional insights regarding the adequacy of your CDISC data for the JMP Clinical built-in reports. Depending on the results of your required variable check, you might need to make some modifications to your SDTM data in order to take full advantage of JMP Clinical. In certain situations, however, you might want to modify the requirements of JMP Clinical. In later sections, we show you how to do this, but not before demonstrating some of the features of JMP Clinical.

Safety Analyses

JMP Clinical comes prepackaged with tools for running both standard and cutting-edge safety analyses, some of which are demonstrated here. Many of the graphical displays of safety data touted by Amit, Heiberger, and Lane (2007) can easily be provided by JMP Clinical.

One common concern when the safety of new drugs and biologics is being assessed is that of liver toxicities. The severe consequences of drug-induced liver injury (DILI) were first published by Dr. Hy Zimmerman in 1978. His observations of signs of DILI have since been informally referred to as Hy's Law (Temple, 2001; Reuben, 2004). They are the motivation behind an FDA guidance document titled "Drug-Induced Liver Injury: Premarketing Clinical Evaluation." Observations of bilirubin (BILI) values >2 times the upper limit of normal (ULN) seen concurrently with severe elevations (for example, >3 times the ULN) of a transaminase such as alanine aminotransferase (ALT) or aspartate transaminase (AST) are now typical criteria for identifying cases of meeting Hy's Law and signaling the potential for severe DILI. With a set of criteria such as these applied to standard laboratory data, including standard liver function tests, evaluations of Hy's Law can be applied in a standard manner.

JMP Clinical has a tool for Hy's Law screening under the **Clinical ▶ Findings** menu. In this dialog box, under the **General** tab, is an option to adjust the time lag for identifying Hy's Law cases. By default, elevations meeting the criteria must occur on the same day in order to be classified as meeting the criteria. This is reasonable if visits are spread over long intervals. However, if visits are more frequent, such as daily or weekly, you might improve detection sensitivity by adjusting the sliding scale to allow for a larger time lag between the elevated liver function tests (LFTs).

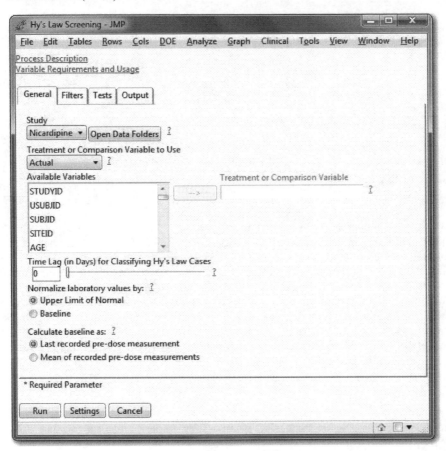

Under the **Filters** tab, more customizations are possible. Although any occurrence of Hy's Law would, in most study populations, be a reason for concern, you can perform the analysis on a chosen set of BY variables or on a particular subset of subjects that can be filtered using the optional WHERE statement. For example, if the ADSL data set contained a flag for subjects with pre-existing liver disease, then this flag could be used to filter these subjects out of the analysis. By default, the analysis is run on the safety population, identified either by the SAFFL flag in ADSL or, if not present, the SAFETY flag in SUPPDM (using the SDTM reserved name).

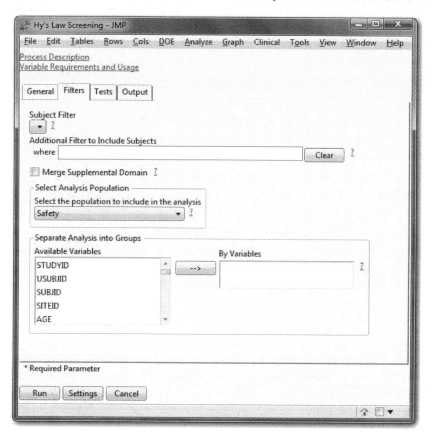

For our example, we ran the Hy's Law Screening routine on the entire nicardipine study population without any BY groups using the default zero time lag. The results from this test are as follows:

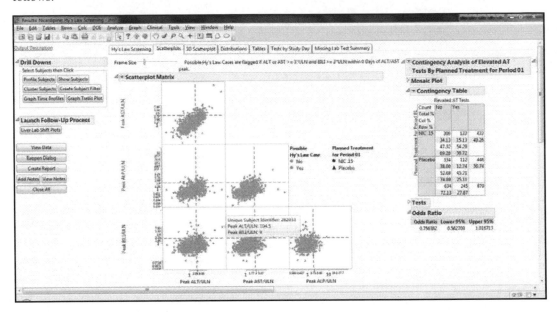

In the scatterplot matrix, the peak value for each relevant lab parameter (AST, ALT, BILI, and alkaline phosphatase, or ALP) is divided by the parameter's upper limit of normal (ULN). These values are then plotted on a log scale (log base 2) for each lab pair. Different shapes are used for each treatment group (in this case, triangles for placebo subjects and circles for nicardipine subjects). Symbols are blue if the subject's data meet the Hy's Law criteria and are red otherwise. The specific criteria are annotated at the top of the scatterplot matrix.

The scatterplot containing BILI on the y-axis and ALT on the x-axis has potential Hy's Law cases in the upper right quadrant. The horizontal reference line for this quadrant is at 2 for BILI (for two times the ULN), and the vertical reference line is at 3 for ALT (for three times the ULN). Note that these reference lines can be changed on the **Output** tab of the Hy's Law Screening dialog box shown previously. Dots that fall within this quadrant are not necessarily classified as patients who meet the criteria because what is being plotted are peak values for each subject. In this way, each subject is represented only once within each scatterplot. Placing your mouse pointer over each point reveals the subject IDs and the lab values divided by the ULN. Two blue dots, signaling nicardipine subjects that meet the criteria, are outliers in each of the plots. These two subjects are 141058 and 282031. By placing your pointer over the dot for subject 282031, we can see that this subject's peak ALT and BILI values are 104 times and 9 times the ULN, respectively. A contingency table in the right pane displays the number of days that subjects spent meeting the criteria.

Next to the **Scatterplots** tab in the center panel is a tab labeled **3D Scatterplot**, which displays results for three of the lab parameters at once: ALT, AST, and BILI. Subjects 141058 and 282031 again appear as outliers, this time with respect to each of the three LFTs. Selecting one of the dots enables you to take advantage of the options under **Drill Downs** in the left pane, which are **Profile Subjects**, **Show Subjects**, **Cluster Subjects**, **Create Subject Filter**, **Graph Time Profiles**, and **Graph Trellis Plot**. Patient profiles are addressed in the next section. Graphing time profiles enables you to view the course of the LFTs over time and to help rule out cases, if any, where the elevations occurred at different times.

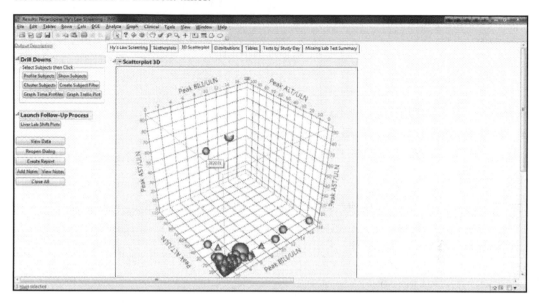

The nicardipine data are especially rich for identifying DILI cases. For other data sets, however, you might want to explore more liberal criteria, either by changing the allowable time lag, or by reducing the ULN multipliers used in the scatterplots. Going back to the Hy's Law Screening dialog box, under the **Options** tab, you can modify the defaults used in the scatterplots, such as the type of log transformation or the reference lines used.

The Hy's Law Screening routine is just one of many tools available in JMP Clinical for safety evaluations. A number of other tools are available for evaluating and summarizing adverse events, laboratory data, and demographics. Like the Hy's Law Screening, many of the others are full of customizations that can make your safety evaluation more relevant to your drug or clinical trial under study. In the next section, we focus on the use of patient profiles for the purpose of examining data across domains for subjects of interest.

Patient Profiles

One of the key features of JMP Clinical is its ability to create patient profiles for safety reviews of individual patients. With your study's CDISC data already registered and the required variable check showing no major problems, you should be able to move ahead with either running profiles for individual patients or, if you want, for the entire study population.

To get started, you must have a subject or set of subjects selected. As shown in the previous section, subjects can be selected from a JMP Clinical graphical display, such as the Hy's Law Screening plots. Another simple way to select a group of subjects is by having a data table open and at least one subject highlighted in that table. When at least one subject has been selected, select **Clinical ▶ Subject Utilities ▶ Profile Subjects** from the Clinical Starter window to generate a report for each subject. The following screenshot shows a sample patient profile for one (subject 282031) of the two nicardipine subjects identified as outliers in the previous section.

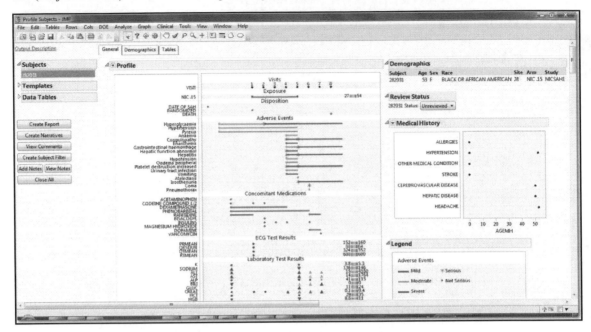

From one profile, clinicians can review all laboratory values, vital signs, ECG results, adverse events, and concomitant medications that a subject experienced during the course of a single study. Baseline information such as demographics and medical history are displayed on the right side. Many times, the automatically generated profiles do not need any alterations in order to get their intended point across. However, for certain studies, or for certain subjects, alterations and customizations might be needed in order to present the salient information in a more readable fashion. Fortunately, the graphical interface provided by JMP gives users the ability to customize the default view in a fairly straightforward point-and-click or click-and-drag manner, both at the domain and term level. These modifications can be saved as a template to be recalled later or provided as the default patient profile view. Users are encouraged to explore these various options as a way to become familiar with the many features that JMP provides.

Whether profiles are being reviewed by a medical monitor during the course of a clinical trial or by a medical reviewer during the course of an NDA review, the clinician might want to add notes regarding his or her findings to each profile. The **Review Status** area in the upper right panel of each profile provides an area for notes to be added and a mechanism for marking each profile as reviewed.

An added feature for those who may want to read or access the profiles without requiring access to JMP (or even a digital device for that matter) is the ability to export the profiles to a separate file that can be shared electronically or as a hard copy. To do this, click the **Create Report** button on the left panel. Provide a name for the file; select the type (**PDF** or **RTF**); adjust the sizes; select a subject or group of subjects for whom you would like profiles created and whether you would like to add medical history; and review comments, the legend, and the data tables that have been added to the patient profile document. (Note that the data tables will be sent to a separate file.) Getting the profile to fit properly on a page might take some trial and error, but hopefully when one set of settings is found to work for one subject, those settings can be applied to all subjects for whom you want paper reports.

Event Narratives

One of the newer features in JMP Clinical, and possibly one of the most useful, is its ability to automatically create safety event narratives. Safety event narratives are typically labor-intensive, manually written write-ups that describe the clinical circumstances surrounding a particular safety event, such as a serious adverse event (SAE). They describe the demographics and medical history of the subject who experienced the event; exposure information to the clinical trial study drug; the events surrounding the event, such as other AEs, lab values, and vital signs; and, of course, the event itself. Historically, patient profiles have been useful for providing these data to the person responsible for writing the narrative or separate data listings, but these have also historically required additional man-hours. With standardized data and some clever, artificial-intelligence-like algorithms, these man-hours can be replaced with computer automation. (However, be aware that some man-hours are usually recommended for quality control, and possible customizations that may be necessary beyond what can be performed automatically.)

In order to avoid confusion, keep in mind that in a clinical trial and pharmacovigilance setting, there are typically two types of safety narratives. SAE narratives are those that typically accompany an SAE report and are usually written by the investigator who witnessed or has first-hand knowledge of the event. SAE reports have certain required information contained within them and sometimes this information, particularly the information contained within the written SAE narrative, is not captured within the study's CRFs.

Often for regulatory submissions and Clinical Study Reports (CSRs), additional narratives are needed not only for SAEs, but also for events that lead to study drug discontinuation or possibly deaths that may not have been considered adverse events (for example, deaths in a cancer trial that either occur well beyond the last dose of study drug or were not unexpected due to disease progression). This latter type is the type that can be generated by JMP Clinical.

Getting back to our example patient profile for subject 282031, notice the button on the left panel for **Create Narratives**. This will open up the narrative dialog box that can also be accessed from the menu via **Clinical ▶ Events ▶ Adverse Events ▶ AE Narrative**. Selecting this sequence of options from the patient profile will prepopulate the **Additional Filter to Include Subjects** field shown on the **Filters** tab. Also, on this tab, we will manually add AESEV="SEVERE" to the **Filter to Include Adverse Events** field as shown in the screenshot. This is added so that the narrative reports include only the severe events that this subject experienced.

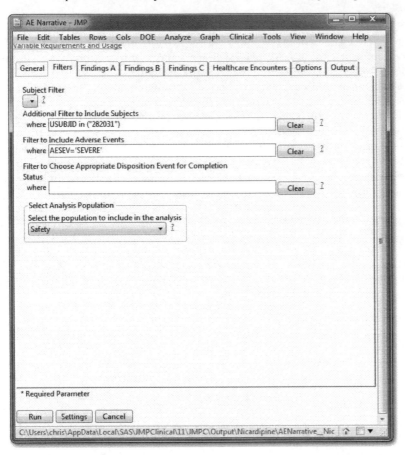

As you can see, there are many possible customizations that can be made to the narrative. We won't go through all of them, but let's look at one example, the tab under **Findings A**. This one defaults to the laboratory data that appear in the narrative (but as can be seen on the drop-down menu, any findings domain can be selected).

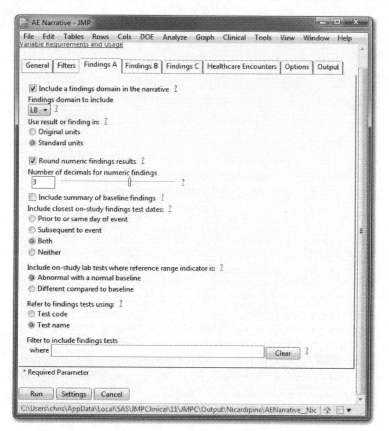

Deciding which options you like best may take some trial and error. It may depend, for example, on the number of relevant lab results that appear and whether you consider the patient's baseline status relevant. For this example, we have chosen, in the interest of space, not to include a summary of baseline results. Since this patient was identified originally from the Hy's Law Screening report, we could set filters to include only the severe or serious hepatic-related events and then filter the labs so that only liver function tests are reported.

After you click Run, the algorithms churn for a few seconds, and up comes an RTF file (presumably opened in Microsoft Word on your machine) with a narrative for each severe AE experienced by subject 282031. In the screenshot, we have selected the one that is related to the severe liver toxicities identified by the Hy's Law Screening.

Subject: 282031
Randomized Arm: NIC .15
Investigator Name: 282A
Drugs and Doses on Day of Event: On Treatment

Serious Adverse Event (coded term [reported term]): HEPATIC FUNCTION
ABNORMAL [HEPATIC FUNCTION ABNORMAL]

Subject 282031 was a 53-year-old black or African American female. Her medical history
included headache associated with sah (1988), hypertension with this sah (1988),
cerebrovascular disease (1986), hepatic disease (1986), allergies (start date unknown),
hypertension prior to sah (start date unknown), other medical condition (start date unknown), and
stroke (start date unknown). The subject discontinued the trial on 05AUG1988 (Day 8) due to
death.

On 01AUG1988 (Day 4) the subject experienced a hepatic function abnormal (severe) which
was considered a serious adverse event (SAE). Though the event was considered serious, no
reasons were provided on the case report form. The subject was on treatment when the event
occurred. It is not known from the case report form if therapeutic measures were administered to
treat the event.

Adverse events that occurred within a +/- 3-day window of the onset of the SAE included
anaemia (moderate), atelectasis (moderate), coagulopathy (severe), coma (severe), enanthema
(mild), gastrointestinal haemorrhage (moderate), gastrointestinal haemorrhage (severe), hepatitis
(severe), hyperglycaemia (severe), hypotension (moderate), isosthenuria (severe), oedema
peripheral (moderate), platelet destruction increased (severe), pneumothorax (moderate), urinary
tract infection (moderate), and vomiting (mild). Concomitant medications taken at the onset of
the SAE included: dexamethasone, insulins, and phenobarbital.

On the closest laboratory test results day on or prior to the start of the event (01AUG1988, Day
4), the subject had the following abnormal on-study laboratory test results: high blood urea
nitrogen [8.925 mmol/L, range = (2.499 - 7.497)], high creatinine [0.133 mmol/L, range = (0.057
- 0.106)], and low platelet [54 U/L, range = (100 - 500)]. On the closest laboratory test results

---

As you read through these narratives, your initial reaction may be amazement that they were
written by a machine. As you read closer, you may notice some idiosyncrasies here and there,
such as abbreviations that are not capitalized that probably should be (such as frequent references
to "sah" for Subarachnoid Hemorrhage in the medical history) or perhaps items that are
capitalized that don't need to be (such as vital sign units written as "BEATS/MIN"), but these are
minor criticisms that can either easily be ignored or fixed with a simple search-and-replace.

Although you should take caution to ensure that the information contained within the narratives is
in fact correct, especially if they are to be included in a regulatory submission, it shouldn't take
long to appreciate the advantages that such a tool can provide compared to the expensive, time-
consuming, and error-prone manual method.

One PROC Away with ADaM Data Sets

Since safety analyses and safety data tend to be more standard across studies, without regard to
the therapeutic area or drug class, they lend themselves well to the development of a tool that can
perform those standard tasks. Efficacy data, however, are rather different. Although the statistical
methods used to analyze efficacy data are somewhat discrete, the number of endpoints, the
number of analysis visits, the different approaches for dealing with missing values, and the
different analysis populations create a myriad of methods and techniques for summarizing and
analyzing efficacy data. This makes standard tool development a bit more complicated.

A basic tenet of ADaM data, as mentioned in Chapter 1, is that the data be both traceable and
ready for analysis. To make the data ready for analysis, flags and other variables are often used in
a WHERE statement to filter the data down to the records of interest. Then one SAS procedure
with this aforementioned WHERE statement (or a WHERE clause) should be all you need to run
the analysis.

This brings us to another basic tenet of ADaM data—that it be accompanied by metadata. One particular piece of metadata (the programming statements that can accompany analysis results metadata) can be used to show the exact programming statements necessary to replicate an analysis (or to run an analysis for the first time). If such metadata are not provided, then an analysis plan should suffice, provided it adequately describes how to conduct the analysis.

So, although the many different types of endpoints and analyses associated with efficacy data make tool construction less straightforward, the process for running an analysis after the ADaM data and metadata have been created is rather simple:

1. Read the documentation (either the analysis results metadata or the analysis plan).
2. Filter down the data.
3. Run a statistical procedure on it.

In order to demonstrate this rather straightforward process, consider two analyses of the pain data from our sample data in earlier chapters: a responder analysis by visit and a time-to-pain relief analysis. The CRIT1FL variable in ADEF can be used for the responder analysis by visit. Suppose responder rates between the treatment groups are to be compared by use of Fisher's exact test. The following code conducts this analysis:

```
*----------------------------------------------------------------*;
* Use Fisher's exact test to compare responder rates
* between treatment groups by visit
*----------------------------------------------------------------*;
PROC SORT
  DATA = ADAM.ADEF
  OUT = ADEF;
    BY CRIT1 AVISITN;
RUN;

ODS SELECT CrossTabFreqs FishersExact;
PROC FREQ
  DATA = ADEF;
    BY CRIT1 AVISITN;
    WHERE ITTFL='Y';
    TABLES TRTPN * CRIT1FL / CHISQ;

  TITLE "Fishers Exact Test on Responder Rates Between Treatment
Groups by Visit";
RUN;
```

Technically, having to sort the data makes the analysis two PROCs away, but most agree that the sort does not count. Alternatively, you could select just one set of records for analysis and therefore not need the BY statement. In this example, the only necessary filtering involves keeping the subjects in the ITT population. Other common scenarios would involve additional filtering, perhaps by using analysis flags (for example, selecting records where ANLzzFL='Y').

Next, we will look at the time-to-event analysis using PROC LIFETEST. We use the following code for this example.

```
PROC LIFETEST
  DATA = ADTTE PLOTS=s;
    WHERE ITTFL = 'Y' and PARAMCD = 'TTPNRELF';
```

```
        ID USUBJID;
        STRATA TRTPN;
        TIME AVAL*CNSR (1, 2, 3);
        TEST TRTPN;
   RUN;
```

From this one procedure, we can produce Kaplan Meier estimates, have them plotted for each treatment group, and conduct a log-rank test.

So, unlike analyses of safety data, which tend to be fairly standard across studies and therapeutic areas, efficacy analyses are more specific from one study to the next. This can make tool construction more difficult. However, with the proper metadata (such as programming statements for analysis results metadata) and properly constructed ADaM efficacy data (that are "one PROC away"), running key analyses can be rather simple. In fact, it is even conceivable that a tool could be developed to automate this process. We leave that as a challenge for the reader.

Transposing the Basic Data Structure for Data Review

As discussed in earlier chapters, the BDS data structure is very flexible for a wide range of applications. But dating back to when it was first being developed and tested as a usable standard for FDA review purposes, there were concerns among some that it didn't totally support their review needs. Indeed, the focus of an FDA reviewer does differ from the focus of an analysis programmer. Sponsor programmers and statisticians tend to be primarily interested in using ADaM data to create the analysis results for which the data are intended. Secondarily, a sponsor statistician and perhaps some clinicians may wish to use ADaM data for review—peeks at certain patients, subgroups, or other ad hoc collections of records. The FDA reviewer, while interested in using the ADaM data to reproduce the primary and secondary study results, also has an intense interest in, as their title suggests, reviewing the data. How a reviewer goes about this depends on a variety of factors—the therapeutic area, the type of data, even individual reviewer preferences. Obviously, the combinations and permutations are too vast to allow for the development of a data standard that would have broad applicability. As such, the BDS was adopted with the understanding that, for those various situations where a reviewer may prefer some type of data re-organization, in most cases it could be handled as a simple transposition of the BDS. While this is still true, there hasn't been enough done to assist ADaM creators, users, and reviewers alike with this transposition. There is no standard way to go about it operationally. Now that therapeutic area standards have been advancing, the time is right to provide some guidance and tools to assist with this standards gap. In this section, we will review an example of this gap and how a SAS macro can be used to close it.

Progression Free Survival with Investigator and Central Assessment

In oncology clinical trials, especially those for solid tumors, one of the most common efficacy endpoints is progression free survival (PFS). "Progression" in this sense refers to the growth of the tumors under study and is assessed by standard criteria such as the Response Evaluation Criteria in Solid Tumors (RECIST) guidelines. Even with standard criteria by which these assessments are made, progression is still often "in the eyes of the beholder." That is, it is not uncommon for two different reviewers to have conflicting RECIST assessments.

Many oncology trials are open-label, meaning there is no blinding with respect to which arm a patient has been randomized, so the study investigator is aware of who is receiving the control therapy and who is receiving the experimental therapy. To alleviate concerns of bias, an

Independent Review Committee (IRC) is often commissioned to perform a central, blinded assessment of the scans used to determine whether the patients' tumors have progressed. However, even with an IRC in place, the investigator needs to make his or her own assessment of progression during the course of the trial in order to make treatment decisions. For example, if a patient's tumors appear to be progressing, then it is often in the best interest of the patient to try a new course of action sooner rather than later. So many trials that have PFS as a primary endpoint have the primary results based on the IRC assessment with the results from the investigators assessment being considered a sensitivity analysis.

During the course of their review, FDA statisticians and clinicians may be interested in reviewing the degree of concordance, or discordance, between the investigator and IRC assessments of tumor progression or response. To facilitate this, it is often helpful to have the two PFS assessments for the same patient side-by-side in different columns of a data set. From a standard time-to-event BDS structure where the two assessments are represented by different parameters, the records could be transposed. To facilitate this, a SAS macro called **%horizontalize** can be used.

Consider the data set shown in the screenshot. It has the BDS time-to-event (TTE) structure with three parameters being shown for two example patients: PFS (IRC Assessment), PFS (Investigator Assessment), and Overall Survival.

	Unique Subject Identifier (usubjid)	Parameter Code (PARAMCD)	Parameter (PARAM)	Analysis Value (AVAL)	Time to Event Origin Date for Subject (STARTDT)	Analysis Date (ADT)	Censor (CNSR)	Event or Censoring Description (EVNTDESC)
1	XYZ123-101-1001	PFS	PFS (IRC Assessment) (days)	303	03NOV2014	02SEP2015	0	Progression
2	XYZ123-101-1001	PFS_PI	PFS (Investigator Assessment) (days)	420	03NOV2014	28DEC2015	1	Last Assessment
3	XYZ123-101-1001	OS	Overall Survival (days)	420	03NOV2014	28DEC2015	1	Last Assessment
4	XYZ123-101-1002	PFS	PFS (IRC Assessment) (days)	299	06NOV2014	30AUG2015	0	Progression
5	XYZ123-101-1002	PFS_PI	PFS (Investigator Assessment) (days)	416	06NOV2014	25DEC2015	1	Last Assessment
6	XYZ123-101-1002	OS	Overall Survival (days)	416	06NOV2014	25DEC2015	1	Last Assessment

Let's propose that at a pre-NDA meeting, the FDA's clinical reviewer requested that the submission include a data set from the pivotal trial. The data set contains one record per patient and separate columns for the two PFS assessments and overall survival. In order to convey all the relevant information, the censoring indicator, event date, and event description columns for each of the three parameters should also exist as separate columns.

The **%horizontalize** macro was created for a situation such as this (this macro can be found on the author pages). It was developed to transpose parameters, analysis visits, or both from the ADaM basic data structure. Having a tool that can do this in an automated and predictable way avoids the need for standards for similar situations in specific therapeutic areas. Such standards would likely require that the variable names meet the SAS transport file limitation of being only 8 characters in length, which can make the interpretation of the variables based on their names rather limited. The **%horizontalize** macro creates variables that are longer than 8 characters, but that are named in a systematic way—with a prefix that corresponds to the PARAMCD value and/or the AVISITN value and then with a suffix that corresponds to the name of the variable that was transposed (except for the variable that contains the AVAL value, which has no suffix).

An example of using this macro for the given scenario is shown here:

```
data adtte;
  set library.adtte;
    where paramcd in('OS', 'PFS', 'PFS_PI'); ❶
run;
```

```
%horizontalize(indata=adtte,
               outdata=tte,  ❷
               xposeby=paramcd,  ❸
               carryonvars=adt cnsr evntdesc);  ❹
```

❶ The data set is first read in as a WORK data set keeping only the desired parameters for the resulting, transposed data set.

❷ The **outdata** parameter to **%horizontalize** is where the name of the new transposed data set is specified.

❸ The **xposeby** parameter can take on values of either PARAMCD, AVISITN, PARAMCD AVISITN, or AVISITN PARAMCD. Those are the only 4 allowable options. This parameter is used to indicate by which variables the data set should be transposed.

❹ The **carryonvars** parameter is used to specify which variables are to be transposed along with AVAL (or AVALC). This can often include variables like ADT (the default value) and ADY. In the example above, we have involved the TTE-specific variables CNSR and EVNTDESC in addition to ADT.

A fourth parameter, **sortby**, is not shown. It is used to specify the sort order and key variables in the resulting data set. The default value is USUBJID, which works for this example, so it therefore does not need to be changed.

The resulting data set should have one row per **sortby** variable and additional columns for every PARAMCD/AVISITN combination (containing the value of AVAL or AVALC), plus additional columns for every PARAMCD/AVISITN/**carryonvar** combination. Variables AVAL, AVALC, PARAMCD/PARAM and/or AVISITN/AVISIT, and each **carryonvar** are dropped.

	Unique Subject Identifier (usubjid)	Time to Event Origin Date for Subject (STARTDT)	Overall Survival (days) (OS)	Overall Survival (days)/Analysis Date (OS_adt)	Overall Survival (days)/Censor (OS_cnsr)	Overall Survival (days)/Event or Censoring Description (OS_evntdesc)	PFS (IRC Assessment) (PFS)	PFS (IRC Assessment) (days) (PFS)	PFS (IRC Assessment) (days)/Censor (PFS_cnsr)	PFS (IRC Assessment) (days)/Event or Censoring Description (PFS_evntdesc)	PFS (Investigator Assessment) (days) (PFS_PI)	PFS (Investigator Assessment) (days)/Analysis Date (PFS_PI_adt)	PFS (Investigator Assessment) (days)/Censor (PFS_PI_cnsr)	PFS (Investigator Assessment) (days)/Event or Censoring Description (PFS_PI_evntdesc)
1	XYZ123-101-1001	03NOV201	420	28DEC201	1	Last Assessme	303	02SEP201	0	Progression	420	28DEC2015	1	Last Assessment
2	XYZ123-101-1002	06NOV201	416	25DEC201	1	Last Assessme	299	30AUG201	0	Progression	416	25DEC2015	1	Last Assessment

As expected, we have two rows for each of the two subjects in the example and 3 x (1+3) = 12 new columns of transposed data that now enables a reviewer to compare PFS days, dates, and censoring indicators for the two types of PFS assessments plus similar additional information about overall survival.

In this example, there were no visits involved. However, as stated above, AVISITN can also be added as an **xposeby** variable. Doing this adds values of AVISITN as the prefix to variable names (preceded by an underscore if it is the only **xposeby** variable since variable names cannot start with a numeric value). This is similar to how PARAMCD values are added as variable name prefixes when PARAMCD is the **xposeby** variable.

Having this standard approach to transposing BDS-structured ADaM data sets (and naming the transposed variables) has several advantages. One is that it obviates the need for the transposed data sets to be created by sponsors. This, in turn, means that the transposed data sets do not have to abide by SAS transport file variable name limitations. It also prevents the need for these transposed data sets to follow ADaM principles and standards. A validated tool such as SAS macro that can implement the standard approach can provide easy access to the transposed data for both sponsors and reviewers. With the development of therapeutic area standards, there is a possibility that these ad hoc transposed data structures created by sponsors for marketing applications could instead be replaced by standard parameter-naming conventions and variable-naming conventions that could be applied to transposed data sets created "on the fly" with review tools.

Chapter Summary

The promise of data standards has long been built on the assumption that they lead to the development of software and other tools to make routine tasks more automatic. JMP Clinical is an example of this predicted evolution. When applied to CDISC data, JMP Clinical can do numerous routine safety analyses very easily. It can do many advanced safety analyses as well. And, like all JMP products, it is especially adept with graphical data displays. Some of these features were demonstrated in this chapter.

The "one-PROC-away" tenet of ADaM data also simplifies data analysis and supports the possibility of tool development. Examples were shown in order to demonstrate the dividends that adapting to ADaM standards pays.

The flexibility of the BDS structure lends itself to the creation of transposed data sets for side-by-side review purposes. The **%horizontalize** macro was introduced as a way that standard variable names (and labels) could be applied to data sets that are transposed by parameters, visits, or both.

Chapter 11: Integrated Data and Regulatory Submissions

The term *integration* in clinical trial parlance has its roots in the Code of Federal Regulations that outlines the content and format of a new drug application. These regulations state that applicants shall submit an "integrated summary of the data demonstrating substantial evidence of effectiveness for the claimed indications" and "an integrated summary of all available information about the safety of the drug product." The terms *integrated summary of safety* and *integrated summary of efficacy* came about in subsequent guidance documents to describe these two cornerstones of the NDA submission. In order to meet the needs of SAS programmers responsible for placing key study results side-by-side in pages of output, or to present adverse event rates with patients pooled and accounted for in one denominator, the patient-level data also had to be combined, or integrated, into one database.

On the surface, the concept of integrating CDISC data seems straightforward. In fact, one of the primary motivations behind data standards has been to facilitate data integration. In concept, with similar (or identical) formats across studies, like data could be stacked on top of one another using a study identifier to distinguish between studies and unique subject IDs to differentiate data. For example, Subject 101 in Study 301 would be differentiated from Subject 101 in Study 302. Not only would this make life easier for the programmers working on integration for a regulatory submission, but it would also make life easier for statistical and medical reviewers at the FDA who are responsible for investigating, in great detail, safety signals that appeared to be part of a drug class effect. Ideally, data from various sponsors, who at one point in time all had their own proprietary data format, could be combined, stored, and extracted from one common database.

However, when the people working on data standards started delving into the details, they began to realize that, in practice, data integration could be a little more complicated. In this chapter, we discuss some of these challenges with integrating data, strategies for data integration, and tools for preparing your CDISC data for an FDA submission.

In the absence of any direct advice from regulators, an ADaM team has been organized to help tackle the problem of how to integrate standardized data for regulatory submissions. They have released a draft version of a document titled "CDISC ADaM Data Structures for Integration: General Considerations and Model for Integrated ADSL (IADSL)." This document will be referenced throughout this chapter.

See Also

CDISC ADaM Data Structures for Integration: General Considerations and Model for Integrated ADSL (IADSL), Version 1.0 Draft (http://www.cdisc.org/standards/foundational/adam)

ADaM Structure for Occurrence Data (OCCDS) version 1.0 (http://www.cdisc.org/standards/foundational/adam)

Study Data Technical Conformance Guide (http://www.fda.gov/downloads/ForIndustry/DataStandards/StudyDataStandards/UCM384744.pdf)

Regulatory Guidance

An abundance of regulatory guidance addresses integrated summaries and data pooling. Here is a list of some of the documents that contain this guidance:

- ICH M4E(R1)–Clinical Overview and Clinical Summary of Module 2
- FDA, Guidance for Industry: Integrated Summary of Effectiveness
- FDA, Guidance for Industry: Premarketing Risk Assessment
- FDA, Study Data Technical Conformance Guide
- EMA, Points to Consider on Application with 1. Meta-Analyses; 2. One Pivotal Study

Much of the language in the first two documents comes from the original 1988 FDA Guideline for the Format and Content of the Clinical and Statistical Sections of an Application (a.k.a. the "Clin-Stat guidance"). These documents provide important information about how and when to conduct integrated analyses and meta-analyses. However, they do not go into detail about how to put an integrated database together.

It should be pointed out that the scope of an integrated database can go beyond the integrated analyses. An integrated database might contain studies that are not necessarily valid for pooling or summarizing with other studies in a regulatory submission, but might be worthwhile for other purposes, such as an organization's pharmacovigilance responsibilities.

Data Integration Challenges

As previously mentioned, the process of integrating data is not always clear-cut. For example, consider a key efficacy variable that collects subject assessments of improvements from baseline on a Likert scale. One study could have categories for 1=None, 2=A Little Better, 3=Better, 4=A

Lot Better, and 5=Completely Better (No Symptoms at All). Another very similar study could have categories for -1 = Worse, 0 = No Improvement, 1=Somewhat Improved, 2=Improved, and 3=Greatly Improved. Not only is the wording of the categories different, but so are the numeric ratings, making the task of presenting the data in a unified way more challenging. How does this happen? Sometimes there are good intentions to make improvements from earlier studies. Sometimes the changes are at the behest of regulatory agencies. Sometimes changes are made to reflect new standards adapted by working groups. Whatever the reason, the question about how to deal with the change often is not asked until the integration work needs to be done.

In other cases, the challenges have less to do with the data collection and more to do with the study design. Consider open-label extension trials for example. These are studies where, for the sake of capturing long-term safety data, subjects being treated for a chronic condition are rolled over from a double-blind, randomized trial that has a primary objective of capturing efficacy data; these subjects are rolled over to an open-label extension trial where everyone is given the experimental treatment and followed primarily for long-term safety evaluations. In rarer cases, subjects are enrolled in extension trials and just observed for long-term safety or efficacy, without any continued experimental intervention. One challenge that these studies present is how to represent study subjects who appear in multiple trials in an integrated summary and in a subject-level data set such as DM (for SDTM data) or ADSL (for ADaM data).

With regard to the subject's unique identifier, the SDTM IG and the CDER Common Issues document are very clear that USUBJID should be unique for any individual within a submission. It is therefore the responsibility of the sponsor, and SAS programmers involved with the studies, to ensure that study subjects who are rolled over from one study to the next are tracked and that only one ID is used for these subjects across studies. A common convention is to construct USUBJID as a concatenation of the study ID, the site ID, and the subject ID (or, if the subject ID already contains the site ID, just the study and subject IDs). For subjects who appear in multiple studies, the ID of the first study in which the subject was enrolled could be used for the USUBJID throughout.

One other challenge when integrating data is deciding whether you should "up version" your data to a newer standard. The need to up version could occur when different studies are based on different versions of the same standard or when individual studies are in the same version of the same standard, but a newer version exists that might be preferred. In particular, if this problem relates to your SDTM data, then the consequences of your decision on the ADaM data also have to be given consideration. In such cases, a change to your SDTM data might necessitate a change to your ADaM data and metadata as well.

Data Integration Strategies

With regard to how to represent subjects who appear in multiple studies in subject-level integrated data sets (such as DM and ADSL), the solution is rarely obvious and not yet prescribed by CDISC standards or regulatory guidance. The options range from representing subjects on multiple rows (one row for each study), representing subjects only once (regardless of the number of previous studies), or representing subjects using some combination of these two options. Throughout this chapter, when speaking of this situation, we are primarily considering the scenario where subjects are rolled over from an earlier trial to an extension trial. Having the same subject appear in multiple unrelated trials (for one development program) is not commonly allowed for a clinical development program. Even if it is allowed, identifying such subjects is often difficult for the programmer or others involved with trial operations. This is due to a combination of privacy regulations that restrict the amount of personal data that can be collected and a lack of a CRF field, or other supporting data, to indicate which subjects have appeared in previous trials.

Another integration strategy that deserves some thought is whether to first integrate the SDTM data and then, from there, integrate the ADaM data, or whether to integrate only the ADaM data from the ADaM data sets in the individual studies.

The advantages and disadvantages of these different approaches are discussed in the following sections.

Representing Subjects Who Appear in Multiple Studies in Subject-Level Data Sets

A common approach to dealing with the integration of subject-level data sets is to keep one record per subject per study, even if this means subjects who appear in multiple studies are represented in multiple rows. This approach is typically easier than the alternative of collapsing subjects who appear in multiple studies into one row of data. It was also supported, to some extent, by the former CDER Common Data Standards Issues document, which stated (before being supplanted by the Study Data Technical Conformance Guide):

> "Integrated summaries may contain more than one record per unique subject in the case that an individual subject was enrolled in more than one study."

This pragmatic recommendation was a motivation behind a subsequent recommendation by the ADaM integration sub-team to allow the same subject to be represented in multiple rows in IADSL, if required by analysis needs.

One drawback of this approach is that it can complicate counting of unique subjects, particularly if pre-existing macros, functions, or software were built under the assumption that each subject is represented only once in data sets such as DM and ADSL. For example, in an integrated summary of adverse events, subjects who appear in multiple studies should be counted only once in denominators when calculating adverse event rates. This is because these denominators represent the number of subjects at risk for an event. If a data set contains the same subject on multiple rows and the AE reporting macro does nothing to check or correct for this assumption, then the results would be incorrect. The IADSL structure includes a flag variable, UADSLFL, that can be used for reducing the data set to one record per subject.

Another issue to consider is that subjects can change treatment groups as they go from one study to the next. Then some thought has to go into how to summarize subjects by treatment group. Take the earlier example of the randomized trial that is followed by an open-label extension trial. In the open-label extension, everybody receives the experimental treatment. Subjects who were initially randomized to receive a placebo will therefore switch treatments from placebo to the experimental treatment. Table 11.1 displays sample data for such a scenario.

Table 11.1: DM Data for a Double-Blind Study and Open-Label Extension Study

STUDYID	USUBJID	ARMCD
Study 1:		
XYZ123	XYZ123-UNI101	PLACEBO
XYZ123	XYZ123-UNI102	ALG123
XYZ123	XYZ123-UNI103	PLACEBO
Study 1 Extension:		
XYZ124	XYZ123-UNI101	ALG123
XYZ124	XYZ123-UNI102	ALG123

Different organizations can implement different strategies for such a scenario. They might summarize AEs by individual treatment group so that subjects who receive different treatments are represented multiple times—once for each unique treatment that they receive. Or they might summarize AEs by a treatment sequence, as is done in crossover trials. Subjects who cross over from the placebo to the experimental treatment would be represented in one summary group. An example of how to represent this in an IADSL data set, using one record per subject per study, is shown in Table 11.2. The reserved variable TRTSEQP, which exists in the ADaM IG specifically for representing a subject's planned treatment sequence, is used to pool treatment sequences across trials.

Table 11.2: Integration Option 1–IADSL Data for a Double-Blind Study and Open-Label Extension Study (One Record per Subject per Study)

STUDIES	USUBJID	ANLCAT	NUMSTUDY	ARMCD	TRT01P	TRTSEQP
Study 1:						
XYZ123	XYZ123-UNI101	DB	2	PLACEBO	Placebo	Placebo/ Analgezia
XYZ123	XYZ123-UNI102	DB	2	ALG123	Analgezia HCL 30 mg	Analgezia/ Analgezia
XYZ123	XYZ123-UNI103	DB	1	PLACEBO	Placebo	Placebo/ Analgezia
Study 1 Extension:						
XYZ124	XYZ123-UNI101	EXT	2	ALG123	Analgezia HCL 30 mg	Placebo/ Analgezia
XYZ124	XYZ123-UNI102	EXT	2	ALG123	Analgezia HCL 30 mg	Analgezia/ Analgezia
Study 1 and the Extension Study:						
XYZ123+ XYZ124	XYZ123-UNI101	DB+EXT	2	PLACEBO	Placebo	Placebo/ Analgezia
XYZ123+ XYZ124	XYZ123-UNI102	DB+EXT	2	ALG123	Analgezia HCL 30 mg	Analgezia/ Analgezia

Note subject XYZ123-UNI103, who did not roll-over into the extension trial. This subject's TRTSEQP value is `Placebo/Analgezia`, even though she never enrolled in the extension trial. For such subjects, you could then use TRTSEQA to represent the actual treatment sequence, which, for this subject, could contain a value of `Placebo` or `Placebo Only`.

Note also the use of three new variables that appear in the IADSL draft document: STUDIES replaces STUDYID to allow multiple studies to be concatenated, thereby indicating the record to be used for integrated or pooled summaries; NUMSTUDY is an integer that indicates the number of studies a subject participated in; and ANLCAT is used to define an analysis category associated with each record. A fourth new variable mentioned earlier, UADSLFL, is not shown but is required to be populated with a Y for one and only one record per unique subject.

The approach of thinking of extension trials in similar terms to the crossover trial provides the motivation for constructing the integrated ADSL data set in a way similar to how an ADSL data set would be constructed for a crossover trial—with one record per subject, but multiple columns for each treatment received. Because ADSL was designed with crossover studies in mind (and other study designs that allow subjects to receive multiple treatment groups), this is rather straightforward. Table 11.3 displays an example for such a scenario.

Table 11.3: Integration Option 2–ADSL Data for a Double-Blind Study and Open-Label Extension Study Combined (One Record per Subject)

STUDYID	STUDYID2	USUBJID	ARMCD	TRT01P	TRT02P	TRTSEQP
XYZ123	XYZ124	XYZ123-UNI101	PLACEBO	Placebo	Analgezia HCL 30 mg	Placebo/Analgezia
XYZ123	XYZ124	XYZ123-UNI102	ALG123	Analgezia HCL 30 mg	Analgezia HCL 30 mg	Analgezia/Analgezia
XYZ123		XYZ123-UNI103	PLACEBO	Placebo		Placebo/Analgezia

In this example, rows of data for subjects who entered the extension trial are replaced with new columns such as STUDYID2 and TRT02P. STUDYID2 is not in the ADaM IG. It was added to capture additional study IDs for situations such as this. TRT02P does exist in the ADaM IG. Here again, the variable TRTSEQA could be used to display the actual treatment sequence for patients such as XYZ123-UNI103 who do not enter the extension trial.

There are many other columns that could be added to capture information specific to the extension trial. Examples include the enrollment date, flags to indicate which subjects entered the extension trial, and treatment start and stop dates (TR02SDT and TR02EDT).

For this set of studies, this integration solution might work. There can be other studies involved with the integration that can complicate matters. Consider another study similar to Study 1 but without an extension (we will call it Study 2). The trick then becomes how to combine treatment groups. For example, should placebo subjects from Study 2 be grouped and summarized with placebo subjects from Study 1 who switch to active drug in the extension phase? When it comes to integration, the scenarios, and methods for dealing with them, can be quite complicated. As the saying goes, the devil is in the details.

Deciding Which Data to Integrate

When you are integrating ADaM data only, the integrated analysis data sets are created from the ADaM data that was created for the individual studies. The primary advantage of this approach is that, by not having to integrate the SDTM data, it involves less work. There are some issues to consider, however, when taking this route:

- *There is no source SDTM data*: As emphasized throughout this book, one of the underlying assumptions of ADaM data (and one of the ADaM principles) is that there is corresponding SDTM data to which the ADaM data can be traced. When you are taking the approach of integrating ADaM data only, the source data (or immediate predecessor) becomes the ADaM data from the individual studies, which could create some traceability problems. For example, you will need to decide how to populate variables that support traceability such as SRCDOM. Should it refer to the same analysis data set at the study level, or should it refer to the SDTM domain at the study level? If referring to the same analysis data set at the study level, then is an additional field needed to identify the source study? Or will this be obvious from the STUDYID?
- With integrated SDTM data, the obvious choice for ADaM traceability variables would be to refer to the integrated SDTM data.
- Certain ADaM data might not exist at the study level: Not all summaries done for a clinical study report necessarily have a corresponding ADaM data set created. For many

standard yet simple summaries, such as a summary of medical history, there are few derivations and less of a need to create ADaM data. Rather than conduct another set of transformations, programmers might instead want to simply merge in the treatment codes and produce summaries using the SDTM data. Not only is this easier for programmers, but it also gives data reviewers one less data format with which to become familiar.

- *Analysis data that do not exist at the study level might be needed at the integration level:* When taking the ADaM-only integration approach, you might find yourself with a need to summarize data that do not exist in ADaM at the individual study level across studies. You might therefore end up having to integrate SDTM data anyway in order to create the required integrated analysis data.

The problems noted above with the ADaM-only integration approach can be avoided by instead integrating the SDTM data first. When you are deciding which approach to take, these points must be weighed against the time and effort required to integrate the SDTM data. Any other potential benefits gained from integrating SDTM data must also be given careful consideration, especially in the context of how medical reviewers, both at regulatory agencies and with sponsor companies, evaluate safety.

At the study level, reviewers rely on SDTM data for two primary reasons: 1) to gain insight into the data's lineage from collection to analysis and 2) to have access to all data collected (which is particularly important for a clinical review). With respect to the latter reason, sometimes certain data that are not used for analysis are needed after analyses are conducted. This can be for reasons that are unforeseen when preparing an analysis plan. At the submission level, the same thing is often true. Reviewers will need access to data that are not summarized in a report, but that are perhaps needed for some other reason.

Consider a medical reviewer who wants to look into the background information for all subjects in a submission who experienced AEs with fatal outcomes. This can involve medical and disease history and other data elements that are not typically provided in analysis data sets. Without such data integrated in SDTM data sets, the reviewer must manually pull the data from the individual studies, which can be much more time-consuming compared to having all of the data already available in one location.

Coding Dictionary Issues

Another common issue with integrated data is dealing with the various versions of coding dictionaries that were used across the individual trials. With MedDRA, for example, which is updated every six months, sometimes just keeping the version consistent within a study can be a challenge. With regard to integrated data, this issue is also recognized in the FDA's Study Data Technical Conformance Guide, which states:

> "Regardless of the specific versions used for individual studies, pooled analyses of coded terms across multiple studies (e.g., for an integrated summary of safety) should be conducted using a single version of a terminology. This will ensure a consistent and coherent comparison of clinical and scientific concepts across multiple studies . . . ."

For MedDRA codes, it is possible to programmatically update coding from an older version to a more recent version. With access to the original lower-level term (LLT), programmers can merge the old LLT with the current version of the MedDRA dictionary to find the new mapping from the LLT to the preferred term (PT). Then, with the new PT, mappings to the higher levels of the MedDRA hierarchy can also be updated. The other alternative is to re-code all events from scratch. This might be necessary if new LLTs exist that better represent the investigator's description of the event. In practice, depending on the number of events, a combination of these two approaches might be most appropriate.

When events are re-coded, it is important to maintain traceability to the original coding used in earlier reports. For example, if a study report for a pivotal study contains events that do not appear in an integrated summary, a reviewer might question the integrity of the data. This is not an uncommon issue when coding is updated.

In an integrated SDTM AE data set, the coding variables should reflect the new coding. To see what changed from the original study, reviewers must manually make comparisons.

However, in an integrated ADAE data set, the original coding can exist on the same row of data that contains the new coding. The variables that allow you to trace back to the original coding are shown in Table 11.4.

Table 11.4: Historic Coding Variables Recommended by ADAE

Variable Name	Variable Label
DECDORGw	PT in Original Dictionary w
BDSYORGw	SOC in Original Dictionary w
HLGTORGw	HLGT in Original Dictionary w
HLTORGw	HLT in Original Dictionary w
LLTORGw	LLT in Original Dictionary w
LLTNORGw	LLT Code in Original Dictionary w

The suffix w represents an integer (1-9) that corresponds to a previous version. The metadata should include the dictionary name and version for each variable. For example, consider a study in which one report is done for an interim analysis and another is done for the final analysis. Assume that this study also appears in an integrated analysis. Also assume that version 10.0 of MedDRA was used for the interim analysis, version 12.1 was used for the final analysis, and version 15.0 was used for the integrated analysis. In the integrated ADAE data set, DECDORG1 could contain the preferred term coding used for the interim analysis, and DECDORG2 could contain the preferred term coding used for the final analysis. AEDECOD would contain the term coding used for the integrated analysis.

The metadata for all three variables should describe the version of MedDRA used for each of these variables. However, this can be problematic if multiple studies have multiple coding in an integrated data set. DECDORG1 might contain version 10.0 coding for one study, but version 8.1 for a different study. When many studies are being integrated and they all have used different versions of the dictionary in their original analysis, the metadata would have to be quite extensive to describe this history for all studies.

Summary of Data Integration Strategies

Ultimately, the decision on how to integrate clinical trial data requires careful planning and consideration toward the issues pointed out above and in regulatory guidance documents. Many of the decisions depend on the number of trials for a particular development program, the design of those trials, and the number of different treatment groups evaluated in those trials. In summary, the list of possibilities might not be endless, but enumerating them all goes well beyond the scope of this book.

Lastly, before implementing any major decisions for an FDA submission, be sure to discuss your issues and proposals with your review division beforehand.

Data Integration and Submission Tools

Similar to the conversion process itself from raw or collected data to the SDTM and from the SDTM to ADaM, data integration can be automated only to a certain extent. Issues described earlier in this chapter need to be addressed, for the most part, on a case-by-case manner. For the actual integration work, of course, the SAS CDI tool can be used in a fashion similar to that illustrated in Chapter 5. Assuming that the data being integrated all have similar (or identical) CDISC structures, the integration process will be much easier than the conversion process shown in Chapter 5.

However, that is not to say that tools cannot be developed to assist with expected common tasks. Such tools are discussed in the following sections. One tool can be used to address the issue of large file sizes in integrated data. Another addresses conversions from native SAS data sets to the version 5 transport format needed for FDA submissions (and conversions back from the transport format to native SAS).

Setting Variable Lengths Based on the Longest Observed Value

Both the CDER Common Issues document and the FDA's Study Data Specifications address an issue that can often occur with integrated data, especially lab data—data sets that are so big they are difficult to open and work with on a standard-issue PC or laptop. As stated in the Study Data Conformance Guide:

> "The allotted length for each column containing character (text) data should be set to the maximum length of the variable used across all datasets in the study. This will significantly reduce file sizes. For example, if USUBJID has a maximum length of 18, the USUBJID's column size should be set to 18, not 200."

The habit of assigning character variable lengths to 200 is tempting because it avoids the need to investigate beforehand how long a field actually needs to be to avoid truncation. In some cases, such as with --TESTCD variables, the length is predefined (to eight characters in the case of --TESTCD variables). In many other cases, however, there is no such restriction (aside from it being less than 200 characters due to the SAS transport file limitation).

In order to avoid the problem of unnecessarily inflating data set file sizes, a macro could be used to determine the maximum observed length of character fields and to then dynamically assign the length based on the data. This does, admittedly, go against the virtue that we have espoused of defining your metadata (including variable lengths) up-front. But during the integration process, when the studies to be integrated tend to be complete or mostly complete, this can be evaluated beforehand. Alternatively, there could be a loop-back process where the metadata are updated after the integration.

The following macro performs this task. The macro is called MAXLENGTH.

```
/*--------------------------------------------------------------------
This macro determines the minimum required length of a variable,
    based on the maximum length observed
```

DATALIST should be a list of similar data sets (for example, from
      different studies) that each contain each variable in
      VARLIST, separated by a space

VARLIST should be a list of character variables for which the
      maximum length is to be examined each variable in the list
      should be separated by a space

Both DATALIST and VARLIST can contain only one item

Set INTEGRATE=1 if you want to have all datasets in DATALIST
   combined into one data set via: SET &DATALIST

If INTEGRATE=1, then IDS should contain the name of the resulting
   Integrated Data Set

```
-------------------------------------------------------------------*/

%macro maxlength(datalist, varlist, integrate=0, ids= );

  ** create global macro variables that will contain the
  **    maximum required length for each variable in &VARLIST;
  %let wrd = 1;
  %do %while(%scan(&varlist,&wrd)^= );
     %global %scan(&varlist,&wrd)max ;
     %let wrd = %eval(&wrd+1);
  %end;

  ** initialize each maximum length to 1;
  %do %while(%scan(&varlist,&wrd)^= );
    %let %scan(&varlist,&wrd)max=1; ❶
    %let wrd = %eval(&wrd+1);
  %end;

  ** find the maximum required length across each data ❷
  ** set in &DATALIST;
  %let d = 1;
  %do %while(%scan(&datalist,&d, )^= );
    %let data=%scan(&datalist,&d, );
    %put data=&data;

    %let wrd = 1;
    data _null_;
      set &data end=eof;

      %do %while(%scan(&varlist,&wrd)^= );
        %let thisvar = %scan(&varlist,&wrd) ;
        retain &thisvar.max &&&thisvar.max ;
        &thisvar.max = max(&thisvar.max,length(&thisvar));
        if eof then
          call symput("&thisvar.max", put(&thisvar.max,4.));
        %let wrd = %eval(&wrd+1);
      %end;
    run;
```

```
         %let d = %eval(&d+1);
    %end;

    %let datasets=%eval(&d - 1);
    %if (&integrate=1 and &ids^= ) or &datasets=1 %then
      %do;

        %let wrd = 1;
        data %if &integrate=1 %then &ids; %else &datalist; ;
          length %do %while(%scan(&varlist,&wrd)^= );
                   %let thisvar=%scan(&varlist,&wrd);
                   &thisvar    $&&&thisvar.max..
                   %let wrd = %eval(&wrd+1);
                 %end;
          ;
          set &datalist;
        run;
      %end;
    %else %if &integrate=1 and &ids= %then
      %put PROBLEM: Integration requested but parameter IDS is blank;
    ;

%mend maxlength;
```

❶   Each variable in &VARLIST becomes a global macro variable with MAX appended to the end of the macro variable name. These macro variables are initialized to a value of 1.

❷   In this DO-WHILE loop, each data set specified in &DATALIST is evaluated separately. The values of the --MAX macro variables are compared to the largest observed length for each new data set and, if necessary, are re-assigned to the new largest observed value.

In order to test the macro, consider the following sample program:

```
data a b c;
  length x y z $200 ;

        x = 'CDISC, SDTM, ADaM';
        y = 'Y';
        z = 'xxxxxxxxxxxxxxxxxxxxxxxxxxxxxxxxxxxxxxx';
        output;
        x = 'hi';
        output;
run;

%maxlength(work.a work.b c, x y z, integrate=1, ids=d);

%put xmax=&xmax ymax=&ymax zmax=&zmax;
proc contents
  data = d;
run;
```

The maximum lengths of variables X, Y, and Z are evaluated in each data set A, B, and C. With the parameters INTEGRATE=1 and IDS=D, these three data sets are SET together to create integrated data set D. Before the SET statement and after the DATA statement, a LENGTH statement appears that assigns the lengths of variables X, Y, and Z using the &XMAX, &YMAX,

and &ZMAX global macro parameters. The LOG file shows the following results from the %put command:

```
%put xmax=&xmax ymax=&ymax zmax=&zmax;
xmax=  17 ymax=   1 zmax=  40
```

The output file shows that the lengths of variables X, Y, and Z in data set D are, as expected, 17, 1, and 40, respectively. Applied to a large data set with a lot of padded character variables, the result can be a significant reduction in the size of the data set.

There is one potential problem to be aware of with this approach, however. Keep in mind that certain variables appear in multiple data sets. Having different lengths for these variables in these different data sets could cause truncation problems when merging data sets. To address this problem, another macro can be used that looks for the minimum required length to avoid truncation across all data sets in the same directory. This macro, called %MAXLENGTH2, is a bit more complicated and too long to include within the text of this book, but it can be found online on the authors' SAS web pages (http://support.sas.com/publishing/authors/index.html). An example of how this macro could be used in practice appears in a later section of this chapter titled "Getting Submission Ready."

Converting from Native SAS to Version 5.0 Transport Files

Many programmers might prefer to keep their SDTM and ADaM data sets in a native SAS format until they are ready to submit the data to the FDA for a regulatory submission. SAS provides a set of macros that facilitate conversions from native SAS formats to SAS transport files. These can be found at http://www.sas.com/en_us/industry/government/accessibility/accessibility-compliance.html#fda-submission-standards. As suggested by the URL, these macros are to help programmers convert their data for FDA submissions.

Many of these macros build off one another. The %TOEXP macro converts a library of native SAS data sets to a separate directory of XPT files. It calls another macro, %DIRDELIM, which is also included in the set. The %TOEXP macro is quite simple to use:

```
%let outdir=&path\data\sdtm\xpt2;
%let indir=&path\data\sdtm;
%toexp(&indir, &outdir);
```

These macros are designed to work in a Microsoft Windows (or PC) or a UNIX or Linux environment. They can handle format catalogs as well (although that is not needed when dealing with SDTM and ADaM data sets because they should have no user-defined formats attached).

Because validation software such as Pinnacle 21 Community requires data sets to be transport files, this macro can also be used before running another software application's validation report. (See Chapters 8 and 9.)

Converting from Version 5.0 Transport Files to Native SAS

Regulatory reviewers and, perhaps, anyone else who prefers to deal with native SAS data sets rather than transport files, might be more interested in the %FROMEXP macro. As the name suggests, this macro converts a directory of transport files to native SAS data sets. Coupling this macro with the %MERGSUPP macro, reviewers can quickly convert their submission data to native SAS data sets and have all supplemental qualifiers merged in with their parent domains, which can make review work much easier.

Here is a sample demonstration:

```
%let indir=&path\data\sdtm\xpt2;
%let outdir=&path\data\sdtm\mergsupp;
%fromexp(&indir, &outdir);

libname sdtm "&outdir";
%mergsupp(sourcelib=sdtm, outlib=sdtm);
```

Running these macros on the SDTM data that were created in earlier chapters creates a version of the DM domain (in the MERGSUPP subdirectory) that has columns for RANDDTC and RACEOTH, the two supplemental qualifiers in SUPPDM.

Getting Submission Ready

Assume that you have finished creating your SDTM and ADaM data, have used them to run analyses for a clinical study report, and are ready to submit them for a new drug application. Now you can use tools that you have learned about in this and other chapters to create and run a simple little program to achieve three important tasks:

1. Convert the data sets from their native SAS format to the SAS transport file format.
2. Shorten text strings to the minimum required length (directory-wide) to avoid truncation (per FDA and CDER recommendations).
3. Run a Pinnacle 21 Community report on the final data.

The following program can achieve these tasks with a minimal amount of code:

```
%let sdtmpath=&path\data\sdtm;
%let truncpath=&sdtmpath\trunc;
%let xptpath=&path\data\sdtm\xpt;
libname sdtm "&sdtmpath";
libname trunc "&truncpath";

%maxlength2(sourcelib=sdtm, outlib=trunc);   ❶
%toexp(&truncpath, &xptpath);   ❷
%run_p21 (sources=&xptpath,config=config-sdtm-
3.1.2.xml,define=N);❸
```

❶ The %MAXLENGTH2 macro was mentioned in the earlier section ("Setting Variable Lengths Based on the Longest Observed Value"), but was not shown due to its length. (But it is available on the authors' web pages at http://support.sas.com/publishing/authors/index.html.) This macro reads in all of the character variables from all of the data sets in the directory specified by the SOURCELIB parameter and determines the minimum length required to avoid truncation (with the exceptions of **TESTCD or PARAMCD variables, which are set to a length of $8; **TEST variables, which are set to a length of $40; and **DTC variables, which are set to a length of $20). Numeric variables are assigned a length of 8. The macro writes the new data sets with the shortened text fields to the directory specified by the OUTLIB parameter. Remember that changing the lengths of variables will affect your metadata. As a result, some sort of loop-back is needed so that those length values can be corrected.

❷ The macro %TOEXP was just introduced in a previous section titled "Converting from Native SAS to Version 5.0 Transport Files." It converts the data sets to XPT files.

❸ Finally, using the %run_p21 macro introduced in Chapter 9, the newly created transport files are checked for their SDTM compliance. As specified by that macro, results are written to an Excel file in the same directory where the data sets reside.

Chapter Summary

Ease of integration is one of the primary motivations for developing data standards. However, even data with common formats and structures across multiple studies present some challenges when it comes to integration. Considerations for dealing with some of these issues are discussed in this chapter. The process of doing the actual integrations can be a manual one, with Base SAS programming, or performed using SAS CDI, as shown in Chapter 5.

Certain macros were introduced in this chapter to assist with data integration and regulatory submissions (and review). The %maxlength macro can be used to determine the minimum length needed to avoid truncation of character fields without unnecessary padding that can greatly inflate data set sizes. Other macros available from the SAS website were shown to demonstrate how you can easily convert an entire directory of native SAS data sets to transport files and, for reviewers, how you can convert an entire directory of transport files back to native SAS data sets, while also using the %mergsupp macro (introduced in Chapter 7) to merge supplemental qualifiers into their parent domain along the way.

Chapter 12: Other Topics

This book has primarily focused on the use of SAS products for the creation and use of data that conforms to CDISC standards, specifically the SDTM and ADaM. In this chapter, we introduce you to other CDISC standards and related initiatives that have less of a direct impact on the typical roles of data managers, SAS programmers, and biostatisticians involved with clinical research. As a result, these standards and initiatives might have little to do with SAS software, at least for now. In the future, however, they could become important topics for understanding the flow of data from outside sources to clinical trials, and vice versa.

Standard for Exchange of Non-Clinical Data

The Standard for Exchange of Non-Clinical Data, or SEND, is detailed in version 3.0 of the SEND Implementation Guide (SENDIG), which was released in May 2011. (See http://www.cdisc.org/standards/foundational/send.) The SENDIG is intended to guide the organization, structure, and format of standard non-clinical tabulation data sets for interchange between organizations such as sponsors and CROs and for submission to the FDA. The current version of the IG is designed to support single-dose general toxicology, repeat-dose general toxicology, and carcinogenicity studies. For non-clinical animal data, the SEND standard is closely related to the SDTM. It contains many of the same domains used for clinical data, in addition to ones specific to non-clinical studies such as death diagnosis (DD), body weight (BW) and body weight gain (BG), and food and water consumption (FW).

Because of the similarities between SEND data and the SDTM, the conversion process from raw data to the SEND standard would presumably be similar to the processes outlined in Chapters 3, 4, and 5, with a similar up-front attention to metadata outlined in Chapter 2. The FDA's study data specifications document has information about acceptance and expectations of the SEND standard in regulatory submissions.

Dataset-XML

Dataset-XML is a relatively new CDISC standard that was developed by the CDISC XML Technologies team as an exchange format for study data sets. Like the Define-XML CDISC standard, the Dataset-XML model is based on the ODM, as depicted in the following figure.

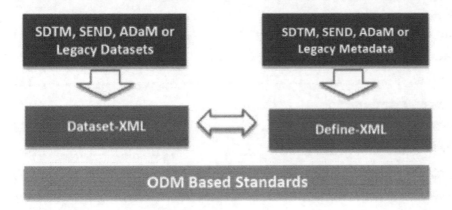

Source: http://www.cdisc.org/dataset-xml.

The idea behind Dataset-XML is for it to serve as an alternative exchange format to the Version 5 SAS Transport (XPT) file format for data sets. It has the advantage of fewer limitations with respect to variable name and label lengths, data set name and label lengths, and text field lengths. In 2014, FDA published a Federal Register Notice requesting participation in a pilot project to evaluate CDISC Dataset XML. FDA envisioned several pilot projects to evaluate new transport formats. As part of the Dataset XML pilot, FDA requested sponsors to prepare and submit previously submitted study data sets using the Dataset XML transport format. The testing period ended in November 2014, and a report was posted online on April 8, 2015 (http://www.fda.gov/downloads/ForIndustry/DataStandards/StudyDataStandards/UCM443327.pdf). The results of the pilot were encouraging, although one limitation discovered during the testing was that of substantial file size inflation when converting data sets to XML from XPT (one sponsor's data set increased 269%, from <5 gigabytes to >17 gigabyes).

FDA envisions conducting several pilots to evaluate new transport formats before a decision is made to support a new format. In the meantime, there are already tools that can be used to convert XPT files to the Dataset-XML format, including some SAS tools. (These tools are summarized on the CDISC wiki at http://wiki.cdisc.org/display/PUB/CDISC+Dataset-XML+Resources.)

Version 1.7 of the Clinical Standards Toolkit support Dataset-XML, but this is a later version than that used for this book. SAS also provides free macros to support the conversion Dataset-XML directly from SAS data sets, which can circumvent the need to create XPT files beforehand. However, an accompanying define.xml file is required (since the define.xml contains information about the data that Dataset-XML does not, such as data set labels). You can learn more about these macros at http://support.sas.com/kb/53/447.html.

There is another tool covered in this book that supports Dataset-XML. Pinnacle 21 Community can convert your XPT files to XML with a straightforward interface.

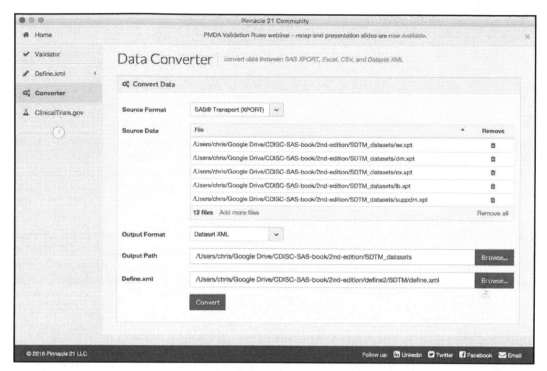

Whether SAS programmers are involved in this transformation process is unclear at the moment. Ideally, the transformation from CDISC-formatted SAS data sets to an approved XML format will be handled by a simple application such as a SAS procedure or something similar that can make the process as seamless as that from native SAS to the version 5 transport file format.

BRIDG Model

BRIDG is an acronym that stands for the Biomedical Research Integrated Domain Group. You can study this model in depth at http://bridgmodel.org. The BRIDG model was conceived by CDISC in 2003 as a way to not only integrate, or bridge, the CDISC models with HL7, but also as a way to harmonize the various internal and disparate CDISC models. BRIDG today is intended to serve as an informational bridge between CDISC, HL7, FDA, and the National Cancer Institute at NIH and its Cancer Biomedical Informatics Grid (caBIG). The main target of BRIDG is to achieve semantic harmonization of information for protocol-driven research across these organizations.

The mechanism of harmonization that BRIDG uses is called a Domain Analysis Model or DAM. A DAM in BRIDG is a general area of interest specified in both Unified Modeling Language (UML) as well as the HL7 Reference Information Model (RIM). For example, the CDISC SDTM and CDASH have been modeled into the BRIDG via DAM processes. You can download the current BRIDG model from their website and see via UML how those models are mapped into the BRIDG and by extension how those models relate to HL7 and the NCI caBIG models.

Over the past two years, a group of individuals has been working on building a BRIDG Statistics DAM. The idea is to map the processes that we perform in clinical trials analysis and reporting and to express that in UML. As part of that effort, the CDISC ADaM model has been translated and mapped into the overarching BRIDG model as well. The hope is that the BRIDG Statistics DAM will enable CDISC, HL7, NCI, and FDA to speak about clinical trials statistics with the same semantic understanding.

Protocol Representation Model

Continuing with the BRIDG initiative, CDISC has released standards that support the BRIDG model. An example of this is the Protocol Representation Model (PRM). (See http://www.cdisc.org/standards/foundational/protocol.) The PRM was developed to capture the elements of a clinical trial protocol such as entry criteria, study objectives, information about study sites, planned activities, and the requirements for ClinicalTrials.gov and World Health Organization (WHO) registries. Much of the SDTM trial design data, which we introduced in Chapter 2, has been mapped to the PRM. In the future, this should take the burden off SAS programmers as the providers of such data. As a follow-up to the PRM itself, CDISC released a PRM toolset. The initial release of this toolset contained two tools: a template for study outline documents and a template for the study outline concept list. The study outline template is a straightforward Word document that can be used to capture all of the key elements of a study. The study outline concept list is an Excel file that contains each study outline concept along with a description, notes, and a mapping of the concept to an element in the SDTM Trial Summary (TS) domain. A desired next step is to create a web tool that can capture each concept and, in turn, create output files such as a PDF version of the study outline, SDTM domain data sets (such as TS and TI), and corresponding ODM XML files.

Some third-party organizations are developing their own template-driven software that can create a Microsoft Word version of a protocol using PRM data elements. The success of such systems should accelerate the uptake of this standard within the clinical research industry.

FDA Janus Clinical Trials Repository

In 2003, the FDA began the task of building a clinical data warehouse called the Janus Clinical Trial Repository (CTR) (http://www.fda.gov/ForIndustry/DataStandards/StudyDataStandards/ucm155327.htm). After several and sometimes recursive stages of development, it finally became operational as a pilot at the FDA's White Oak Data Center in January of 2015. As of February 4, 2016, it is in full production mode at the FDA with 172 studies loaded. This is a substantial increase from the previous year when 61 studies had been loaded in an operational pilot. The idea behind Janus is to give FDA personnel who are responsible for approval and surveillance of medical products an integrated clinical data repository that can enhance their ability to carry out the mission of protecting the public health. For example, Janus can be mined to study the comparative effectiveness of interventions or to study the safety of interventions across a class of similar compounds.

It is expected that the FDA requirements to have sponsors electronically submit their marketing applications with standardized (and validated) study data will greatly accelerate the uptake of study data into the CTR, which will ultimately transform how the FDA manages and analyzes clinical trial data and fulfills its mission of protecting the public health.

CDISC Model Versioning

A critical topic worthy of your organization's consideration is how to handle CDISC model versioning. Versioning is the process by which your organization either updates or implements new versions of the CDISC standards. We have not addressed this issue in this book so far, but it is worth serious thought before you implement CDISC models. Some questions that you should be considering are:

- *When will the FDA and your* customers *accept or require the new standard?*

  If you adopt the new CDISC model version, will your customer or the FDA be ready for it? You might not want to adopt the new standard before your end users can use it.

- *How will the new standard impact your internal systems?*

  How will adoption of the new standard impact your internal systems and software? Does your environment allow for multiple standards in production at the same time? How does use of the new standard impact the associated older dependent standards? How will you handle multiple CDISC model versions used within a submission?

As an example of this problem, while producing this book, we encountered the release of "Amendment 1 to the Study Data Tabulation Model (SDTM) v1.2 and the SDTM Implementation Guide: Human Clinical Trials V3.1.2." So we had to decide what to do with it. Recognizing that the CDISC models will constantly be in a state of update and revision and seeing that this particular SDTM amendment did not add much value to the examples in this book, we chose to ignore it for now. You are going to face similar situations in your work, and you and your organization should have governance plans for how to handle the evolving CDISC models.

Future CDISC Directions

In what direction is CDISC heading? There seem to be two paths. One is focused on CDISC's core expertise—standards for clinical trial data (the type that is captured on a CRF). Another is for continued interoperability with other healthcare standards.

Regarding CDISC's core expertise, one hot topic has been the development of SDTM models for specific therapeutic areas. An example of this is the oncology domains that are modeled to capture tumor data. Models have also been created to capture data specific to Alzheimer's disease, tuberculosis, and cardiovascular endpoints. Similar progress is being made with the ADaM standard, mostly through the release of documents that demonstrate examples of how to use existing ADaM models for specific types of analyses, such as time-to-event analysis or categorical data analysis.

Regarding the path of interoperability with other healthcare data standards, CDISC members are actively supporting and developing the BRIDG model (as evidenced by the release of the Protocol Representation Model and the Statistics DAM discussed previously). Another example of this direction is CDISC's Healthcare Link Initiative (see http://www.cdisc.org/healthcare-link), which involves a collaboration with an organization called Integrating the Healthcare Enterprise (IHE). Through this initiative, a method has been established for linking electronic health record data to critical secondary uses such as safety reporting (and biosurveillance), clinical research, and disease registries.

Chapter Summary

The realm of CDISC standards continues to evolve and expand. CDISC standards have expanded to include SDTM domains specific to certain therapeutic areas such as oncology, Alzheimer's disease, and tuberculosis. They have evolved from focusing on standards for clinical trial data intended for an ultimate FDA submission to standards for non-clinical data and clinical trial information that goes beyond what is typically captured on a case report form, such as protocol data.

Despite these evolutions, the roles of SAS programmers, biostatisticians, and data managers are unlikely to be affected to any great degree (at least for the time being). These roles will continue to be instrumental in helping organizations adopt CDISC standards and for the development of tools to make the use of CDISC data more automatic. We do, however, expect to see the tools that we use improve and evolve as the CDISC standards are adopted and implemented across our industry. In the future, we expect to see the development of protocol authoring tools, improved standards-based data collection tools, and statistical computing environments that are based on the efficient use of the CDISC model standards.

Appendix A: Source Data Programs

This appendix presents Base SAS programs that you can run to generate the source data for the SDTM datasets. The trial design dataset source data are available as a Microsoft Excel file called trialdesign.xlsx that can be downloaded from the author page for this book. See support.sas.com/publishing/authors/index.html.

adverse Dataset

```
**** INPUT SAMPLE ADVERSE EVENT DATA.;
data source.adverse;
label subject  = "Subject Number"
      bodysys  = "Body System of Event"
      prefterm = "Preferred Term for Event"
      aerel    = "Relatedness: 1=not,2=possibly,3=probably"
      aesev    = "Severity/Intensity:1=mild,2=moderate,3=severe"
      aeaction = "Action taken: 1=drug stopped, 2=dose reduced, 3=dose increased, 4=no
                  dose change, 5=unknown"
      aestart  = "AE Start date"
      aeend    = "AE End date"
      serious  = "Serious AE?"
      aetext   = "Event Verbatim Text"
      uniqueid = "Company Wide Subject ID";
serious = 'N';
input subject aerel aesev aeaction aestart mmddyy10. +1 aeend mmddyy10. bodysys $ 33-63 prefterm $ 65-82 aetext $ 94-130;
uniqueid = 'UNI' || put(subject,3.);
datalines;
101 1 1 1 04/05/2010 04/05/2010 CARDIAC DISORDERS             ATRIAL FLUTTER              RACING HEART BEAT
101 2 1 2 06/05/2010 06/10/2010 GASTROINTESTINAL DISORDERS    CONSTIPATION               CONSTIPATED
102 2 2 3 02/15/2010 02/15/2010 CARDIAC DISORDERS             CARDIAC FAILURE            CONGESTIVE HEART FAILURE
102 1 1 4 02/15/2010 02/15/2010 PSYCHIATRIC DISORDERS         DELIRIUM                   DISORIENTED AND DELIRIOUS
103 1 1 5 06/05/2010 06/05/2010 CARDIAC DISORDERS             PALPITATIONS               HEART POUNDING
103 1 2 5 07/05/2010 07/05/2010 CARDIAC DISORDERS             PALPITATIONS               HEART POUNDING
103 2 2 2 07/05/2010 07/05/2010 CARDIAC DISORDERS             TACHYCARDIA                ACCELERATED HEARTBEAT
201 3 2 4 07/15/2010 07/25/2010 GASTROINTESTINAL DISORDERS    ABDOMINAL PAIN             STOMACH PAIN
201 3 1 1 10/12/2010 10/12/2010 GASTROINTESTINAL DISORDERS    ANAL ULCER                 ANAL ULCER
205 2 1 4 06/17/2010 06/27/2010 GASTROINTESTINAL DISORDERS    CONSTIPATION               CONSTIPATED
205 2 2 4 04/13/2010 04/20/2010 GASTROINTESTINAL DISORDERS    DYSPEPSIA                  UPSET STOMACH
301 3 3 4 02/21/2010 .          GASTROINTESTINAL DISORDERS    FLATULENCE                 PASSING GAS
302 1 3 2 09/05/2010 09/05/2010 GASTROINTESTINAL DISORDERS    HIATUS HERNIA              HIATAL HERNIA
302 1 1 1 10/01/2010 10/01/2010 NERVOUS SYSTEM DISORDERS      CONVULSION                 CONVULSIONS
303 2 2 2 02/20/2010 02/20/2010 NERVOUS SYSTEM DISORDERS      DIZZINESS                  DIZZY FEELINGS
303 1 1 3 02/19/2010 .          NERVOUS SYSTEM DISORDERS      ESSENTIAL TREMOR           HAND TREMORS
304 1 3 4 05/20/2010 05/25/2010 PSYCHIATRIC DISORDERS         CONFUSIONAL STATE          FELT LOST
304 1 1 5 05/20/2010 05/25/2010 PSYCHIATRIC DISORDERS         DELIRIUM                   DELIRIOUS
304 2 1 1 05/19/2010 .          PSYCHIATRIC DISORDERS         SLEEP DISORDER             COULD NOT SLEEP
305 1 3 4 10/05/2010 10/05/2010 CARDIAC DISORDERS             PALPITATIONS               POUNDING HEART
401 2 2 2 09/09/2010 .          GASTROINTESTINAL DISORDERS    HEARTBURN                  UPSET STOMACH
402 1 1 1 01/05/2010 .          GASTROINTESTINAL DISORDERS    TOOTHACHE                  MOUTH PAIN
402 3 3 4 03/03/2010 03/03/2010 INFECTIONS AND INFESTATIONS   NASOPHARYNGITIS            COMMON COLD
402 2 1 3 06/17/2010 06/17/2010 PSYCHIATRIC DISORDERS         ANXIETY                    ANXIETY
403 1 1 5 04/02/2010 04/02/2010 PSYCHIATRIC DISORDERS         INSOMNIA                   INSOMNIA SYMPTOMS
403 1 1 4 06/10/2010 06/10/2010 PSYCHIATRIC DISORDERS         SUICIDAL IDEATION          SUICIDAL
404 2 2 3 05/01/2010 05/01/2010 SURGICAL AND MEDICAL PROCEDURES EAR OPERATION            RIGHT EAR SURGERY
404 2 2 4 06/10/2010 .          GASTROINTESTINAL DISORDERS    FLATULENCE                 PASSING GAS
405 1 1 4 03/03/2010 .          RENAL AND URINARY DISORDERS   URINARY TRACT OBSTRUCTION  URINARY BLOCKAGE SURGERY
405 1 1 4 07/10/2010 07/10/2010 NERVOUS SYSTEM DISORDERS      SEDATION                   FEELING SLEEPY
501 2 2 1 01/01/2010 01/01/2010 NERVOUS SYSTEM DISORDERS      CARPAL TUNNEL SYNDROME     CARPAL TUNNEL IN LEFT HAND
501 3 2 2 05/20/2010 05/20/2010 GASTROINTESTINAL DISORDERS    VOMITING                   THROWING UP
502 3 2 2 08/10/2010 08/10/2010 GASTROINTESTINAL DISORDERS    ABDOMINAL PAIN             ABDOMINAL CRAMPS
502 1 1 4 10/10/2010 .          EYE DISORDERS                 MYOPIA                     NEARSIGHTED
504 1 1 4 06/19/2010 6/19/2010  EAR AND LABYRINTH DISORDERS   EAR PAIN                   LEFT EAR ACHE
505 1 2 4 03/16/2010 03/16/2010 GASTROINTESTINAL DISORDERS    DIARRHOEA                  LOOSE STOOL
505 2 3 1 09/01/2010 09/01/2010 NERVOUS SYSTEM DISORDERS      HEADACHE                   HEADACHE
505 2 2 4 05/10/2010 05/10/2010 EAR AND LABYRINTH DISORDERS   TINNITUS                   RINGING IN EAR
507 1 1 4 05/05/2010 05/05/2010 GASTROINTESTINAL DISORDERS    IRRITABLE BOWEL SYNDROME   IRRITABLE BOWEL SYNDROME
507 1 2 4 07/07/2010 07/07/2010 NERVOUS SYSTEM DISORDERS      AMNESIA                    MILD MEMORY LOSS
;
run;
```

demographics Dataset

```
proc format;
    value trt
        1 = "Active"
        0 = "Placebo";
    value gender
        1 = "Male"
        2 = "Female";
    value race
        1 = "White"
        2 = "Black"
        3 = "Other";
run;

**** INPUT SAMPLE DEMOGRAPHICS DATA;
data source.demographic;
label subject  = "Subject Number"
      trt      = "Treatment"
      gender   = "Gender"
      race     = "Race"
      orace    = "Oher Race Specify"
      dob      = "Date of Birth"
      uniqueid = "Company Wide Subject ID"
      randdt   = "Randomization Date";
input subject 1-3 trt 5 gender 7 race 9 orace $ 11-20 +1 dob
mmddyy10. +1
                        randdt mmddyy10.;
uniqueid = 'UNI' || put(subject,3.);
format dob randdt mmddyy10.;
datalines;
101 0 1 3 BRAZILIAN   02/05/1974 04/02/2010
102 1 2 1             11/02/1946 02/13/2010
103 1 1 2             05/01/1979 05/16/2010
104 0 2 1             05/01/1972 01/02/2010
105 1 1 3 ABORIGINE   03/02/1979 04/20/2010
106 0 2 1             05/01/1977 04/01/2010
201 1 1 3 LIBYAN      04/28/1949 06/11/2010
202 0 2 1             01/13/1967 02/23/2010
203 1 1 2             01/01/1971 06/10/2010
204 0 2 1             04/17/1950 02/03/2010
205 1 1 3 HMONG       05/13/1978 04/13/2010
206 1 2 1             02/09/1948 07/01/2010
301 0 1 1             04/12/1941 02/20/2010
302 0 1 2             07/02/1978 05/12/2010
303 1 1 1             03/02/1967 02/19/2010
304 0 1 1             03/03/1958 05/19/2010
305 1 1 1             02/28/1966 06/10/2010
306 0 1 2             01/02/1960 05/23/2010
401 1 2 1             10/31/1970 06/13/2010
402 0 2 2             10/12/1980 01/02/2010
403 1 1 1             01/23/1974 03/03/2010
404 1 1 1             03/18/1954 04/24/2010
405 0 2 1             03/19/1939 03/01/2010
406 1 1 2             03/19/1970 06/12/2010
501 0 1 2             06/12/1978 02/23/2010
```

```
502 1 2 1              03/16/1967 05/28/2010
503 1 1 1              08/14/1962 01/21/2010
504 0 1 3 JAPANESE     02/01/1977 06/19/2010
505 1 1 2              04/16/1969 03/13/2010
506 0 1 1              07/22/1964 01/20/2010
507 1 1 2              10/22/1976 04/06/2010
508 0 2 1              12/22/1965 02/04/2010
509 0 1 1              12/13/1972 05/16/2010
510 0 1 3 HMONG        07/24/1956 01/12/2010
511 1 2 2              09/28/1960 04/10/2010
512 1 1 1              03/02/1967 04/03/2010
601 0 1 1              01/12/1961 06/04/2010
602 0 2 2              05/09/1971 02/28/2010
603 1 2 1              07/14/1980 01/12/2010
604 0 1 1              11/11/1977 03/13/2010
605 1 2 1              12/31/1955 04/23/2010
606 0 1 2              09/20/1971 02/22/2010
607 1 1 1              10/13/1966 06/10/2010
608 0 2 2              07/15/1967 02/20/2010
609 1 2 1              10/16/1945 05/22/2010
610 0 1 1              06/22/1954 01/09/2010
611 1 2 1              10/18/1949 06/19/2010
612 1 1 2              02/02/1960 03/12/2010
701 1 1 1              12/02/1951 05/12/2010
702 0 1 1              01/29/1955 02/10/2010
703 1 1 2              02/12/1956 05/20/2010
704 0 2 1              12/25/1972 06/01/2010
705 1 1 2              08/23/1981 05/24/2010
706 1 1 1              12/13/1965 06/12/2010
707 1 1 1              12/12/1972 01/12/2010
708 0 2 1              10/13/1978 03/13/2010
709 0 2 2              10/10/1940 04/23/2010
710 0 1 1              08/11/1949 03/16/2010
711 1 1 2              11/19/1950 04/24/2010
712 0 . 1              12/30/1948 06/10/2010
;
run;
```

dosing Dataset

```
**** INPUT SAMPLE DOSING DATA;
data source.dosing;
label subject  = "Subject Number"
      startdt  = "Dosing start date"
      enddt    = "Dosing end date"
      dailydose= "Daily dose taken (pills)"
      uniqueid = "Company Wide Subject ID"
      startmm  = "Month of Start Dose"
      startdd  = "Day of Start Dose"
      startyy  = "Year of Start Dose"
      endmm    = "Month of End Dose"
      enddd    = "Day of End Dose"
      endyy    = "Year of End Dose";
input subject 1-3 startmm 5-6 startdd 8-9 startyy 11-14
      endmm 16-17 enddd 19-20 endyy 22-25 dailydose 27;
```

```
uniqueid = 'UNI' || put(subject,3.);
startdt = mdy(startmm, startdd , startyy);
enddt = mdy(endmm, enddd, endyy);
format startdt enddt mmddyy10.;
datalines;
101 04/02/2010 07/26/2010 2
101 07/31/2010 10/10/2010 3
102 02/13/2010 03/20/2010 2
102 03/25/2010 08/10/2010 1
103 05/16/2010 11/14/2010 1
104 01/02/2010 01/10/2010 2
104 01/15/2010 05/25/2010 1
104 05/26/2010 07/04/2010 2
105 04/20/2010 07/20/2010 2
105 07/21/2010 10/19/2010 1
106 04/01/2010 10/10/2010 2
201 06/11/2010 12/11/2010 1
202 02/23/2010 05/19/2010 2
203 06/10/2010 06/20/2010 1
204 02/03/2010 05/04/2010 2
204 05/05/2010 08/03/2010 2
205 04/13/2010 10/10/2010 1
206 07/01/2010 10/01/2010 2
206 10/02/2010 12/27/2010 2
301 02/20/2010 05/17/2010 2
301 05/19/2010 08/22/2010 3
302 05/12/2010 08/12/2010 2
302 08/13/2010 11/15/2010 3
303 02/19/2010 08/17/2010 2
304 05/19/2010 11/19/2010 1
305 06/10/2010 12/11/2010 2
306 05/23/2010 08/19/2010 2
306 08/20/2010 11/18/2010 1
401 06/13/2010 12/09/2010 2
402 01/02/2010 01/05/2010 3
402 01/06/2010 04/02/2010 2
402 04/03/2010 07/10/2010 1
403 03/03/2010 09/04/2010 2
404 04/24/2010 10/25/2010 3
405 03/01/2010 08/28/2010 2
406 06/12/2010 12/09/2010 3
501 02/23/2010 08/17/2010 2
502 05/28/2010 08/25/2010 2
502 08/29/2010 11/20/2010 1
503 01/21/2010 07/19/2010 2
504 06/19/2010 12/20/2010 2
505 03/13/2010 06/15/2010 2
505 06/16/2010 06/20/2010 1
506 01/20/2010 07/20/2010 2
507 04/06/2010 10/05/2010 2
508 02/04/2010 08/11/2010 2
509 05/16/2010 11/15/2010 2
510 01/12/2010 07/14/2010 2
511 04/10/2010 10/12/2010 3
512 04/03/2010 10/01/2010 3
601 06/04/2010 12/09/2010 2
602 02/28/2010 03/24/2010 3
```

```
602 03/25/2010 06/22/2010 2
603 01/12/2010 04/12/2010 2
603 04/13/2010 07/10/2010 1
604 03/13/2010 09/15/2010 2
605 04/23/2010 10/26/2010 3
606 02/22/2010 05/25/2010 3
606 05/26/2010 08/23/2010 2
607 06/10/2010 12/09/2010 3
608 02/20/2010 05/20/2010 2
608 05/21/2010 08/17/2010 1
609 05/22/2010 11/23/2010 2
610 01/09/2010 07/12/2010 2
611 06/19/2010 09/21/2010 3
611 09/22/2010 12/20/2010 2
612 03/12/2010 06/15/2010 3
612 06/17/2010 09/16/2010 1
701 05/12/2010 08/10/2010 3
701 08/11/2010 11/05/2010 1
702 02/10/2010 08/12/2010 3
703 05/20/2010 11/20/2010 2
704 06/01/2010 12/10/2010 3
705 05/24/2010 11/19/2010 2
706 06/12/2010 09/11/2010 3
706 09/12/2010 12/10/2010 1
707 01/12/2010 04/11/2010 3
707 04/12/2010 07/10/2010 1
708 03/13/2010 09/15/2010 3
709 04/23/2010 10/23/2010 2
710 03/16/2010 09/16/2010 2
711 04/24/2010 09/20/2010 3
712   /  /2010 09/09/2010 3
712 09/10/2010 12/  /2010 2
;
run;
```

laboratory Dataset

```
**** INPUT SAMPLE LABORATORY DATA;
data source.labs;
label subject     = "Subject Number"
    month       = "Month: 0=baseline, 1=3 months, 2 =6 months"
    labcat      = "Category for Lab Test"
    labtest     = "Laboratory Test"
    colunits    = "Collected Units"
    nresult     = "Numeric Result"
    lownorm     = "Normal Range Lower Limit"
    highnorm    = "Normal Range Upper Limit"
    labdate     = "Date of Lab Test"
    uniqueid = "Company Wide Subject ID";
format labdate mmddyy10.;
input subject 1-3 month 6 labcat $ 9-18 labtest $ 20-30
    colunits $ 32-36 nresult 38-41 lownorm 45-48 highnorm 52-55 +1
    labdate mmddyy10.;
uniqueid = 'UNI' || put(subject,3.);
datalines;
```

```
101   0   HEMATOLOGY  HEMATOCRIT   %      31    35    49     04/02/2010
101   1   HEMATOLOGY  HEMATOCRIT   %      39    35    49     07/03/2010
101   2   HEMATOLOGY  HEMATOCRIT   %      44    35    49     10/10/2010
101   0   HEMATOLOGY  HEMOGLOBIN   g/dL   11.5  11.7  15.9   04/02/2010
101   1   HEMATOLOGY  HEMOGLOBIN   g/dL   13.2  11.7  15.9   07/03/2010
101   2   HEMATOLOGY  HEMOGLOBIN   g/dL   14.3  11.7  15.9   10/10/2010
101   0   CHEMISTRY   AST (SGOT)   IU/L   12    10    34     04/02/2010
101   1   CHEMISTRY   AST (SGOT)   IU/L   40    10    34     07/03/2010
101   2   CHEMISTRY   AST (SGOT)   IU/L   44    10    34     10/10/2010
101   0   CHEMISTRY   ALT (SGPT)   IU/L   10    5     35     04/02/2010
101   1   CHEMISTRY   ALT (SGPT)   IU/L   22    5     35     07/03/2010
101   2   CHEMISTRY   ALT (SGPT)   IU/L   33    5     35     10/10/2010
101   0   CHEMISTRY   ALK. PHOS.   IU/L   33    20    140    04/02/2010
101   1   CHEMISTRY   ALK. PHOS.   IU/L   49    20    140    07/03/2010
101   2   CHEMISTRY   ALK. PHOS.   IU/L   200   20    140    10/10/2010
101   0   CHEMISTRY   GGTP         IU/L   5     0     51     04/02/2010
101   1   CHEMISTRY   GGTP         IU/L   15    0     51     07/03/2010
101   2   CHEMISTRY   GGTP         IU/L   15    0     51     10/10/2010
101   0   CHEMISTRY   DIRECT BILI  mg/dL  0.1   0     0.3    04/02/2010
101   1   CHEMISTRY   DIRECT BILI  mg/dL  0.2   0     0.3    07/03/2010
101   2   CHEMISTRY   DIRECT BILI  mg/dL  0.1   0     0.3    10/10/2010
101   0   CHEMISTRY   TOTAL BILI   mg/dL  1.0   0.3   1.9    04/02/2010
101   1   CHEMISTRY   TOTAL BILI   mg/dL  0.5   0.3   1.9    07/03/2010
101   2   CHEMISTRY   TOTAL BILI   mg/dL  2.5   0.3   1.9    10/10/2010
101   0   CHEMISTRY   ALBUMIN      g/dL   3.3   3.4   5.4    04/02/2010
101   1   CHEMISTRY   ALBUMIN      g/dL   4.1   3.4   5.4    07/03/2010
101   2   CHEMISTRY   ALBUMIN      g/dL   5.5   3.4   5.4    10/10/2010
101   0   CHEMISTRY   TOTAL PROT   g/dL   6.4   6.0   8.3    04/02/2010
101   1   CHEMISTRY   TOTAL PROT   g/dL   7.0   6.0   8.3    07/03/2010
101   2   CHEMISTRY   TOTAL PROT   g/dL   8.2   6.0   8.3    10/10/2010
102   0   HEMATOLOGY  HEMATOCRIT   %      39    35    49     02/13/2010
102   1   HEMATOLOGY  HEMATOCRIT   %      39    35    49     05/10/2010
102   2   HEMATOLOGY  HEMATOCRIT   %      44    35    49     08/11/2010
102   0   HEMATOLOGY  HEMOGLOBIN   g/dL   11.5  11.7  15.9   02/13/2010
102   1   HEMATOLOGY  HEMOGLOBIN   g/dL   13.2  11.7  15.9   05/10/2010
102   2   HEMATOLOGY  HEMOGLOBIN   g/dL   18.3  11.7  15.9   08/11/2010
102   0   CHEMISTRY   AST (SGOT)   IU/L   14    10    34     02/13/2010
102   1   CHEMISTRY   AST (SGOT)   IU/L   45    10    34     05/10/2010
102   2   CHEMISTRY   AST (SGOT)   IU/L   34    10    34     08/11/2010
102   0   CHEMISTRY   ALT (SGPT)   IU/L   10    5     35     02/13/2010
102   1   CHEMISTRY   ALT (SGPT)   IU/L   22    5     35     05/10/2010
102   2   CHEMISTRY   ALT (SGPT)   IU/L   36    5     35     08/11/2010
102   0   CHEMISTRY   ALK. PHOS.   IU/L   33    20    140    02/13/2010
102   1   CHEMISTRY   ALK. PHOS.   IU/L   55    20    140    05/10/2010
102   2   CHEMISTRY   ALK. PHOS.   IU/L   150   20    140    08/11/2010
102   0   CHEMISTRY   GGTP         IU/L   20    0     51     02/13/2010
102   1   CHEMISTRY   GGTP         IU/L   25    0     51     05/10/2010
102   2   CHEMISTRY   GGTP         IU/L   25    0     51     08/11/2010
102   0   CHEMISTRY   DIRECT BILI  mg/dL  0.1   0     0.3    02/13/2010
102   1   CHEMISTRY   DIRECT BILI  mg/dL  0.1   0     0.3    05/10/2010
102   2   CHEMISTRY   DIRECT BILI  mg/dL  0.4   0     0.3    08/11/2010
102   0   CHEMISTRY   TOTAL BILI   mg/dL  1.1   0.3   1.9    02/13/2010
102   1   CHEMISTRY   TOTAL BILI   mg/dL  1.5   0.3   1.9    05/10/2010
102   2   CHEMISTRY   TOTAL BILI   mg/dL  2.5   0.3   1.9    08/11/2010
102   0   CHEMISTRY   ALBUMIN      g/dL   3.3   3.4   5.4    02/13/2010
102   1   CHEMISTRY   ALBUMIN      g/dL   4.1   3.4   5.4    05/10/2010
102   2   CHEMISTRY   ALBUMIN      g/dL   5.2   3.4   5.4    08/11/2010
```

```
102  0  CHEMISTRY   TOTAL PROT   g/dL  6.3   6.0    8.3   02/13/2010
102  1  CHEMISTRY   TOTAL PROT   g/dL  7.1   6.0    8.3   05/10/2010
102  2  CHEMISTRY   TOTAL PROT   g/dL  8.3   6.0    8.3   08/11/2010
103  0  HEMATOLOGY  HEMATOCRIT   %     54    35     49    05/15/2010
103  1  HEMATOLOGY  HEMATOCRIT   %     33    35     49    08/15/2010
103  2  HEMATOLOGY  HEMATOCRIT   %     52    35     49    11/15/2010
103  0  HEMATOLOGY  HEMOGLOBIN   g/dL  12.5  11.7   15.9  05/15/2010
103  1  HEMATOLOGY  HEMOGLOBIN   g/dL  12.2  11.7   15.9  08/15/2010
103  2  HEMATOLOGY  HEMOGLOBIN   g/dL  14.3  11.7   15.9  11/15/2010
103  0  CHEMISTRY   AST (SGOT)   IU/L  19    10     34    05/15/2010
103  1  CHEMISTRY   AST (SGOT)   IU/L  41    10     34    08/15/2010
103  2  CHEMISTRY   AST (SGOT)   IU/L  34    10     34    11/15/2010
103  0  CHEMISTRY   ALT (SGPT)   IU/L  13    5      35    05/15/2010
103  1  CHEMISTRY   ALT (SGPT)   IU/L  22    5      35    08/15/2010
103  2  CHEMISTRY   ALT (SGPT)   IU/L  34    5      35    11/15/2010
103  0  CHEMISTRY   ALK. PHOS.   IU/L  33    20     140   05/15/2010
103  1  CHEMISTRY   ALK. PHOS.   IU/L  120   20     140   08/15/2010
103  2  CHEMISTRY   ALK. PHOS.   IU/L  110   20     140   11/15/2010
103  0  CHEMISTRY   GGTP         IU/L  5     0      51    05/15/2010
103  1  CHEMISTRY   GGTP         IU/L  25    0      51    08/15/2010
103  2  CHEMISTRY   GGTP         IU/L  33    0      51    11/15/2010
103  0  CHEMISTRY   DIRECT BILI  mg/dL 0.1   0      0.3   05/15/2010
103  1  CHEMISTRY   DIRECT BILI  mg/dL 0.1   0      0.3   08/15/2010
103  2  CHEMISTRY   DIRECT BILI  mg/dL 0.2   0      0.3   11/15/2010
103  0  CHEMISTRY   TOTAL BILI   mg/dL 1.1   0.3    1.9   05/15/2010
103  1  CHEMISTRY   TOTAL BILI   mg/dL 1.5   0.3    1.9   08/15/2010
103  2  CHEMISTRY   TOTAL BILI   mg/dL 2.7   0.3    1.9   11/15/2010
103  0  CHEMISTRY   ALBUMIN      g/dL  3.3   3.4    5.4   05/15/2010
103  1  CHEMISTRY   ALBUMIN      g/dL  4.2   3.4    5.4   08/15/2010
103  2  CHEMISTRY   ALBUMIN      g/dL  5.3   3.4    5.4   11/15/2010
103  0  CHEMISTRY   TOTAL PROT   g/dL  6.2   6.0    8.3   05/15/2010
103  1  CHEMISTRY   TOTAL PROT   g/dL  7.1   6.0    8.3   08/15/2010
103  2  CHEMISTRY   TOTAL PROT   g/dL  8.4   6.0    8.3   11/15/2010
104  0  HEMATOLOGY  HEMATOCRIT   %     50    35     49    01/02/2010
104  1  HEMATOLOGY  HEMATOCRIT   %     42    35     49    04/03/2010
104  2  HEMATOLOGY  HEMATOCRIT   %     42    35     49    07/04/2010
104  0  HEMATOLOGY  HEMOGLOBIN   g/dL  13.0  11.7   15.9  01/02/2010
104  1  HEMATOLOGY  HEMOGLOBIN   g/dL  13.3  11.7   15.9  04/03/2010
104  2  HEMATOLOGY  HEMOGLOBIN   g/dL  12.8  11.7   15.9  07/04/2010
104  0  CHEMISTRY   AST (SGOT)   IU/L  22    10     34    01/02/2010
104  1  CHEMISTRY   AST (SGOT)   IU/L  23    10     34    04/03/2010
104  2  CHEMISTRY   AST (SGOT)   IU/L  22    10     34    07/04/2010
104  0  CHEMISTRY   ALT (SGPT)   IU/L  12    5      35    01/02/2010
104  1  CHEMISTRY   ALT (SGPT)   IU/L  22    5      35    04/03/2010
104  2  CHEMISTRY   ALT (SGPT)   IU/L  23    5      35    07/04/2010
104  0  CHEMISTRY   ALK. PHOS.   IU/L  30    20     140   01/02/2010
104  1  CHEMISTRY   ALK. PHOS.   IU/L  37    20     140   04/03/2010
104  2  CHEMISTRY   ALK. PHOS.   IU/L  100   20     140   07/04/2010
104  0  CHEMISTRY   GGTP         IU/L  34    0      51    01/02/2010
104  1  CHEMISTRY   GGTP         IU/L  22    0      51    04/03/2010
104  2  CHEMISTRY   GGTP         IU/L  34    0      51    07/04/2010
104  0  CHEMISTRY   DIRECT BILI  mg/dL 0.1   0      0.3   01/02/2010
104  1  CHEMISTRY   DIRECT BILI  mg/dL 0.1   0      0.3   04/03/2010
104  2  CHEMISTRY   DIRECT BILI  mg/dL 0.1   0      0.3   07/04/2010
104  0  CHEMISTRY   TOTAL BILI   mg/dL 1.3   0.3    1.9   01/02/2010
104  1  CHEMISTRY   TOTAL BILI   mg/dL 1.5   0.3    1.9   04/03/2010
104  2  CHEMISTRY   TOTAL BILI   mg/dL 1.8   0.3    1.9   07/04/2010
```

104	0	CHEMISTRY	ALBUMIN		g/dL	3.6	3.4	5.4	01/02/2010
104	1	CHEMISTRY	ALBUMIN		g/dL	4.3	3.4	5.4	04/03/2010
104	2	CHEMISTRY	ALBUMIN		g/dL	5.3	3.4	5.4	07/04/2010
104	0	CHEMISTRY	TOTAL PROT		g/dL	6.2	6.0	8.3	01/02/2010
104	1	CHEMISTRY	TOTAL PROT		g/dL	7.2	6.0	8.3	04/03/2010
104	2	CHEMISTRY	TOTAL PROT		g/dL	8.3	6.0	8.3	07/04/2010
105	0	HEMATOLOGY	HEMATOCRIT		%	39	35	49	04/20/2010
105	1	HEMATOLOGY	HEMATOCRIT		%	35	35	49	07/20/2010
105	2	HEMATOLOGY	HEMATOCRIT		%	37	35	49	10/19/2010
105	0	HEMATOLOGY	HEMOGLOBIN		g/dL	13.0	11.7	15.9	04/20/2010
105	1	HEMATOLOGY	HEMOGLOBIN		g/dL	14.2	11.7	15.9	07/20/2010
105	2	HEMATOLOGY	HEMOGLOBIN		g/dL	14.6	11.7	15.9	10/19/2010
105	0	CHEMISTRY	AST (SGOT)		IU/L	22	10	34	04/20/2010
105	1	CHEMISTRY	AST (SGOT)		IU/L	23	10	34	07/20/2010
105	2	CHEMISTRY	AST (SGOT)		IU/L	22	10	34	10/19/2010
105	0	CHEMISTRY	ALT (SGPT)		IU/L	12	5	35	04/20/2010
105	1	CHEMISTRY	ALT (SGPT)		IU/L	22	5	35	07/20/2010
105	2	CHEMISTRY	ALT (SGPT)		IU/L	23	5	35	10/19/2010
105	0	CHEMISTRY	ALK. PHOS.		IU/L	30	20	140	04/20/2010
105	1	CHEMISTRY	ALK. PHOS.		IU/L	37	20	140	07/20/2010
105	2	CHEMISTRY	ALK. PHOS.		IU/L	100	20	140	10/19/2010
105	0	CHEMISTRY	GGTP		IU/L	34	0	51	04/20/2010
105	1	CHEMISTRY	GGTP		IU/L	22	0	51	07/20/2010
105	2	CHEMISTRY	GGTP		IU/L	34	0	51	10/19/2010
105	0	CHEMISTRY	DIRECT BILI	mg/dL	0.1	0	0.3	04/20/2010
105	1	CHEMISTRY	DIRECT BILI	mg/dL	0.1	0	0.3	07/20/2010
105	2	CHEMISTRY	DIRECT BILI	mg/dL	0.1	0	0.3	10/19/2010
105	0	CHEMISTRY	TOTAL BILI	mg/dL	1.3	0.3	1.9	04/20/2010
105	1	CHEMISTRY	TOTAL BILI	mg/dL	1.5	0.3	1.9	07/20/2010
105	2	CHEMISTRY	TOTAL BILI	mg/dL	1.8	0.3	1.9	10/19/2010
105	0	CHEMISTRY	ALBUMIN		g/dL	3.6	3.4	5.4	04/20/2010
105	1	CHEMISTRY	ALBUMIN		g/dL	4.3	3.4	5.4	07/20/2010
105	2	CHEMISTRY	ALBUMIN		g/dL	5.3	3.4	5.4	10/19/2010
105	0	CHEMISTRY	TOTAL PROT		g/dL	6.2	6.0	8.3	04/20/2010
105	1	CHEMISTRY	TOTAL PROT		g/dL	7.2	6.0	8.3	07/20/2010
105	2	CHEMISTRY	TOTAL PROT		g/dL	8.3	6.0	8.3	10/19/2010
106	0	HEMATOLOGY	HEMATOCRIT		%	53	35	49	04/01/2010
106	1	HEMATOLOGY	HEMATOCRIT		%	50	35	49	07/05/2010
106	2	HEMATOLOGY	HEMATOCRIT		%	53	35	49	10/10/2010
106	0	HEMATOLOGY	HEMOGLOBIN		g/dL	17.0	11.7	15.9	04/01/2010
106	1	HEMATOLOGY	HEMOGLOBIN		g/dL	12.3	11.7	15.9	07/05/2010
106	2	HEMATOLOGY	HEMOGLOBIN		g/dL	12.9	11.7	15.9	10/10/2010
106	0	CHEMISTRY	AST (SGOT)		IU/L	22	10	34	04/01/2010
106	1	CHEMISTRY	AST (SGOT)		IU/L	33	10	34	07/05/2010
106	2	CHEMISTRY	AST (SGOT)		IU/L	36	10	34	10/10/2010
106	0	CHEMISTRY	ALT (SGPT)		IU/L	13	5	35	04/01/2010
106	1	CHEMISTRY	ALT (SGPT)		IU/L	24	5	35	07/05/2010
106	2	CHEMISTRY	ALT (SGPT)		IU/L	29	5	35	10/10/2010
106	0	CHEMISTRY	ALK. PHOS.		IU/L	30	20	140	04/01/2010
106	1	CHEMISTRY	ALK. PHOS.		IU/L	50	20	140	07/05/2010
106	2	CHEMISTRY	ALK. PHOS.		IU/L	90	20	140	10/10/2010
106	0	CHEMISTRY	GGTP		IU/L	33	0	51	04/01/2010
106	1	CHEMISTRY	GGTP		IU/L	66	0	51	07/05/2010
106	2	CHEMISTRY	GGTP		IU/L	60	0	51	10/10/2010
106	0	CHEMISTRY	DIRECT BILI	mg/dL	0.2	0	0.3	04/01/2010
106	1	CHEMISTRY	DIRECT BILI	mg/dL	0.1	0	0.3	07/05/2010
106	2	CHEMISTRY	DIRECT BILI	mg/dL	0.2	0	0.3	10/10/2010

```
106   0   CHEMISTRY   TOTAL BILI    mg/dL  1.4    0.3    1.9   04/01/2010
106   1   CHEMISTRY   TOTAL BILI    mg/dL  1.6    0.3    1.9   07/05/2010
106   2   CHEMISTRY   TOTAL BILI    mg/dL  1.7    0.3    1.9   10/10/2010
106   0   CHEMISTRY   ALBUMIN       g/dL   3.9    3.4    5.4   04/01/2010
106   1   CHEMISTRY   ALBUMIN       g/dL   4.4    3.4    5.4   07/05/2010
106   2   CHEMISTRY   ALBUMIN       g/dL   5.5    3.4    5.4   10/10/2010
106   0   CHEMISTRY   TOTAL PROT    g/dL   6.0    6.0    8.3   04/01/2010
106   1   CHEMISTRY   TOTAL PROT    g/dL   7.1    6.0    8.3   07/05/2010
106   2   CHEMISTRY   TOTAL PROT    g/dL   8.8    6.0    8.3   10/10/2010
201   0   HEMATOLOGY  HEMATOCRIT    %      48     35     49    06/10/2010
201   1   HEMATOLOGY  HEMATOCRIT    %      59     35     49    09/09/2010
201   2   HEMATOLOGY  HEMATOCRIT    %      60     35     49    12/12/2010
201   0   HEMATOLOGY  HEMOGLOBIN    g/dL   15.0   11.7   15.9  06/10/2010
201   1   HEMATOLOGY  HEMOGLOBIN    g/dL   14.3   11.7   15.9  09/09/2010
201   2   HEMATOLOGY  HEMOGLOBIN    g/dL   19.1   11.7   15.9  12/12/2010
201   0   CHEMISTRY   AST (SGOT)    IU/L   20     10     34    06/10/2010
201   1   CHEMISTRY   AST (SGOT)    IU/L   21     10     34    09/09/2010
201   2   CHEMISTRY   AST (SGOT)    IU/L   20     10     34    12/12/2010
201   0   CHEMISTRY   ALT (SGPT)    IU/L   13     5      35    06/10/2010
201   1   CHEMISTRY   ALT (SGPT)    IU/L   20     5      35    09/09/2010
201   2   CHEMISTRY   ALT (SGPT)    IU/L   39     5      35    12/12/2010
201   0   CHEMISTRY   ALK. PHOS.    IU/L   32     20     140   06/10/2010
201   1   CHEMISTRY   ALK. PHOS.    IU/L   33     20     140   09/09/2010
201   2   CHEMISTRY   ALK. PHOS.    IU/L   111    20     140   12/12/2010
201   0   CHEMISTRY   GGTP          IU/L   32     0      51    06/10/2010
201   1   CHEMISTRY   GGTP          IU/L   23     0      51    09/09/2010
201   2   CHEMISTRY   GGTP          IU/L   39     0      51    12/12/2010
201   0   CHEMISTRY   DIRECT BILI   mg/dL  0.09   0      0.3   06/10/2010
201   1   CHEMISTRY   DIRECT BILI   mg/dL  0.1    0      0.3   09/09/2010
201   2   CHEMISTRY   DIRECT BILI   mg/dL  0.2    0      0.3   12/12/2010
201   0   CHEMISTRY   TOTAL BILI    mg/dL  1.4    0.3    1.9   06/10/2010
201   1   CHEMISTRY   TOTAL BILI    mg/dL  1.7    0.3    1.9   09/09/2010
201   2   CHEMISTRY   TOTAL BILI    mg/dL  1.3    0.3    1.9   12/12/2010
201   0   CHEMISTRY   ALBUMIN       g/dL   3.3    3.4    5.4   06/10/2010
201   1   CHEMISTRY   ALBUMIN       g/dL   4.9    3.4    5.4   09/09/2010
201   2   CHEMISTRY   ALBUMIN       g/dL   5.2    3.4    5.4   12/12/2010
201   0   CHEMISTRY   TOTAL PROT    g/dL   6.3    6.0    8.3   06/10/2010
201   1   CHEMISTRY   TOTAL PROT    g/dL   7.4    6.0    8.3   09/09/2010
201   2   CHEMISTRY   TOTAL PROT    g/dL   8.9    6.0    8.3   12/12/2010
202   0   HEMATOLOGY  HEMATOCRIT    %      41     35     49    01/23/2010
202   1   HEMATOLOGY  HEMATOCRIT    %      52     35     49    04/20/2010
202   2   HEMATOLOGY  HEMATOCRIT    %      48     35     49    07/20/2010
202   0   HEMATOLOGY  HEMOGLOBIN    g/dL   15.1   11.7   15.9  01/23/2010
202   1   HEMATOLOGY  HEMOGLOBIN    g/dL   18.2   11.7   15.9  04/20/2010
202   2   HEMATOLOGY  HEMOGLOBIN    g/dL   17.7   11.7   15.9  07/20/2010
202   0   CHEMISTRY   AST (SGOT)    IU/L   22     10     34    01/23/2010
202   1   CHEMISTRY   AST (SGOT)    IU/L   23     10     34    04/20/2010
202   2   CHEMISTRY   AST (SGOT)    IU/L   22     10     34    07/20/2010
202   0   CHEMISTRY   ALT (SGPT)    IU/L   12     5      35    01/23/2010
202   1   CHEMISTRY   ALT (SGPT)    IU/L   22     5      35    04/20/2010
202   2   CHEMISTRY   ALT (SGPT)    IU/L   23     5      35    07/20/2010
202   0   CHEMISTRY   ALK. PHOS.    IU/L   30     20     140   01/23/2010
202   1   CHEMISTRY   ALK. PHOS.    IU/L   37     20     140   04/20/2010
202   2   CHEMISTRY   ALK. PHOS.    IU/L   100    20     140   07/20/2010
202   0   CHEMISTRY   GGTP          IU/L   34     0      51    01/23/2010
202   1   CHEMISTRY   GGTP          IU/L   22     0      51    04/20/2010
202   2   CHEMISTRY   GGTP          IU/L   34     0      51    07/20/2010
```

202	0	CHEMISTRY	DIRECT BILI	mg/dL	0.1	0	0.3	01/23/2010
202	1	CHEMISTRY	DIRECT BILI	mg/dL	0.1	0	0.3	04/20/2010
202	2	CHEMISTRY	DIRECT BILI	mg/dL	0.1	0	0.3	07/20/2010
202	0	CHEMISTRY	TOTAL BILI	mg/dL	1.3	0.3	1.9	01/23/2010
202	1	CHEMISTRY	TOTAL BILI	mg/dL	1.5	0.3	1.9	04/20/2010
202	2	CHEMISTRY	TOTAL BILI	mg/dL	1.8	0.3	1.9	07/20/2010
202	0	CHEMISTRY	ALBUMIN	g/dL	3.6	3.4	5.4	01/23/2010
202	1	CHEMISTRY	ALBUMIN	g/dL	4.3	3.4	5.4	04/20/2010
202	2	CHEMISTRY	ALBUMIN	g/dL	5.3	3.4	5.4	07/20/2010
202	0	CHEMISTRY	TOTAL PROT	g/dL	6.2	6.0	8.3	01/23/2010
202	1	CHEMISTRY	TOTAL PROT	g/dL	7.2	6.0	8.3	04/20/2010
202	2	CHEMISTRY	TOTAL PROT	g/dL	8.3	6.0	8.3	07/20/2010
203	0	HEMATOLOGY	HEMATOCRIT	%	39	35	49	06/10/2010
203	1	HEMATOLOGY	HEMATOCRIT	%	53	35	49	09/01/2010
203	2	HEMATOLOGY	HEMATOCRIT	%	57	35	49	11/28/2010
203	0	HEMATOLOGY	HEMOGLOBIN	g/dL	13.0	11.7	15.9	06/10/2010
203	1	HEMATOLOGY	HEMOGLOBIN	g/dL	17.3	11.7	15.9	09/01/2010
203	2	HEMATOLOGY	HEMOGLOBIN	g/dL	17.3	11.7	15.9	11/28/2010
203	0	CHEMISTRY	AST (SGOT)	IU/L	17	10	34	06/10/2010
203	1	CHEMISTRY	AST (SGOT)	IU/L	20	10	34	09/01/2010
203	2	CHEMISTRY	AST (SGOT)	IU/L	33	10	34	11/28/2010
203	0	CHEMISTRY	ALT (SGPT)	IU/L	8	5	35	06/10/2010
203	1	CHEMISTRY	ALT (SGPT)	IU/L	12	5	35	09/01/2010
203	2	CHEMISTRY	ALT (SGPT)	IU/L	22	5	35	11/28/2010
203	0	CHEMISTRY	ALK. PHOS.	IU/L	22	20	140	06/10/2010
203	1	CHEMISTRY	ALK. PHOS.	IU/L	33	20	140	09/01/2010
203	2	CHEMISTRY	ALK. PHOS.	IU/L	56	20	140	11/28/2010
203	0	CHEMISTRY	GGTP	IU/L	12	0	51	06/10/2010
203	1	CHEMISTRY	GGTP	IU/L	33	0	51	09/01/2010
203	2	CHEMISTRY	GGTP	IU/L	29	0	51	11/28/2010
203	0	CHEMISTRY	DIRECT BILI	mg/dL	0.1	0	0.3	06/10/2010
203	1	CHEMISTRY	DIRECT BILI	mg/dL	0.2	0	0.3	09/01/2010
203	2	CHEMISTRY	DIRECT BILI	mg/dL	0.3	0	0.3	11/28/2010
203	0	CHEMISTRY	TOTAL BILI	mg/dL	1.0	0.3	1.9	06/10/2010
203	1	CHEMISTRY	TOTAL BILI	mg/dL	0.4	0.3	1.9	09/01/2010
203	2	CHEMISTRY	TOTAL BILI	mg/dL	1.2	0.3	1.9	11/28/2010
203	0	CHEMISTRY	ALBUMIN	g/dL	3.6	3.4	5.4	06/10/2010
203	1	CHEMISTRY	ALBUMIN	g/dL	4.8	3.4	5.4	09/01/2010
203	2	CHEMISTRY	ALBUMIN	g/dL	5.4	3.4	5.4	11/28/2010
203	0	CHEMISTRY	TOTAL PROT	g/dL	6.3	6.0	8.3	06/10/2010
203	1	CHEMISTRY	TOTAL PROT	g/dL	7.4	6.0	8.3	09/01/2010
203	2	CHEMISTRY	TOTAL PROT	g/dL	8.4	6.0	8.3	11/28/2010
204	0	HEMATOLOGY	HEMATOCRIT	%	38	35	49	02/03/2010
204	1	HEMATOLOGY	HEMATOCRIT	%	40	35	49	05/04/2010
204	2	HEMATOLOGY	HEMATOCRIT	%	44	35	49	08/05/2010
204	0	HEMATOLOGY	HEMOGLOBIN	g/dL	11.2	11.7	15.9	02/03/2010
204	1	HEMATOLOGY	HEMOGLOBIN	g/dL	12.2	11.7	15.9	05/04/2010
204	2	HEMATOLOGY	HEMOGLOBIN	g/dL	17.3	11.7	15.9	08/05/2010
204	0	CHEMISTRY	AST (SGOT)	IU/L	20	10	34	02/03/2010
204	1	CHEMISTRY	AST (SGOT)	IU/L	33	10	34	05/04/2010
204	2	CHEMISTRY	AST (SGOT)	IU/L	45	10	34	08/05/2010
204	0	CHEMISTRY	ALT (SGPT)	IU/L	10	5	35	02/03/2010
204	1	CHEMISTRY	ALT (SGPT)	IU/L	20	5	35	05/04/2010
204	2	CHEMISTRY	ALT (SGPT)	IU/L	40	5	35	08/05/2010
204	0	CHEMISTRY	ALK. PHOS.	IU/L	32	20	140	02/03/2010
204	1	CHEMISTRY	ALK. PHOS.	IU/L	33	20	140	05/04/2010
204	2	CHEMISTRY	ALK. PHOS.	IU/L	160	20	140	08/05/2010

```
204  0  CHEMISTRY  GGTP         IU/L  32   0     51   02/03/2010
204  1  CHEMISTRY  GGTP         IU/L  21   0     51   05/04/2010
204  2  CHEMISTRY  GGTP         IU/L  33   0     51   08/05/2010
204  0  CHEMISTRY  DIRECT BILI  mg/dL 0.2  0     0.3  02/03/2010
204  1  CHEMISTRY  DIRECT BILI  mg/dL 0.1  0     0.3  05/04/2010
204  2  CHEMISTRY  DIRECT BILI  mg/dL 0.2  0     0.3  08/05/2010
204  0  CHEMISTRY  TOTAL BILI   mg/dL 1.6  0.3   1.9  02/03/2010
204  1  CHEMISTRY  TOTAL BILI   mg/dL 1.2  0.3   1.9  05/04/2010
204  2  CHEMISTRY  TOTAL BILI   mg/dL 1.8  0.3   1.9  08/05/2010
204  0  CHEMISTRY  ALBUMIN      g/dL  3.6  3.4   5.4  02/03/2010
204  1  CHEMISTRY  ALBUMIN      g/dL  4.2  3.4   5.4  05/04/2010
204  2  CHEMISTRY  ALBUMIN      g/dL  5.1  3.4   5.4  08/05/2010
204  0  CHEMISTRY  TOTAL PROT   g/dL  6.3  6.0   8.3  02/03/2010
204  1  CHEMISTRY  TOTAL PROT   g/dL  7.4  6.0   8.3  05/04/2010
204  2  CHEMISTRY  TOTAL PROT   g/dL  8.2  6.0   8.3  08/05/2010
;
run;
```

pain scores Dataset

```
**** INPUT SAMPLE EFFICACY PAIN DATA;
data source.pain;
label subject  = "Subject Number"
      randomizedt = "Baseline visit date"
      month3dt   = "Month 3 visit date"
      month6dt   = "Month 6 visit date"
      painbase   = "Pain score at baseline: 0=none, 1=mild, 2=moderate,
                      3=severe"
      pain3mo    = "Pain score at 3 months: 0=none, 1=mild, 2=moderate,
                      3=severe"
      pain6mo    = "Pain score at 6 months: 0=none, 1=mild, 2=moderate,
                      3=severe"
      uniqueid = "Company Wide Subject ID";
input subject randomizedt mmddyy10. +1 month3dt mmddyy10. +1
month6dt mmddyy10. painbase pain3mo pain6mo;
uniqueid = 'UNI' || put(subject,3.);
format randomizedt month3dt month6dt mmddyy10.;
datalines;
101 04/02/2010 07/03/2010 10/10/2010 3 2 1
102 02/13/2010 05/10/2010 08/11/2010 3 3 1
103 05/15/2010 08/15/2010 11/15/2010 3 3 0
104 01/02/2010 04/03/2010 07/04/2010 2 0 0
105 04/20/2010 07/20/2010 10/19/2010 3 3 1
106 04/01/2010 07/05/2010 10/10/2010 3 0 0
201 06/10/2010 09/09/2010 12/12/2010 3 2 0
202 01/23/2010 04/20/2010 07/20/2010 3 1 1
203 06/10/2010 .          .          3 . .
204 02/03/2010 05/04/2010 08/05/2010 3 2 1
205 04/13/2010 07/12/2010 10/10/2010 3 2 3
206 07/01/2010 10/01/2010 12/28/2010 2 2 0
301 02/20/2010 05/19/2010 08/22/2010 3 1 1
302 05/12/2010 08/13/2010 11/15/2010 3 3 3
303 02/19/2010 05/17/2010 08/17/2010 3 2 1
304 05/19/2010 08/15/2010 11/20/2010 3 0 0
305 06/10/2010 09/09/2010 12/12/2010 3 3 0
```

```
306 05/23/2010 08/20/2010 11/19/2010 3 0 0
401 06/13/2010 09/11/2010 12/10/2010 3 3 3
402 01/02/2010 04/02/2010 07/10/2010 2 3 3
403 03/03/2010 06/04/2010 09/05/2010 2 3 1
404 04/23/2010 07/24/2010 10/26/2010 2 3 0
405 02/28/2010 05/25/2010 08/28/2010 3 0 0
406 06/12/2010 09/10/2010 12/09/2010 3 3 3
501 02/23/2010 05/20/2010 08/17/2010 3 0 0
502 05/27/2010 08/25/2010 11/20/2010 2 0 0
503 01/19/2010 04/20/2010 07/19/2010 1 0 0
504 06/19/2010 09/20/2010 12/20/2010 2 0 0
505 03/12/2010 06/15/2010 .          3 0 .
506 01/20/2010 04/19/2010 07/20/2010 3 0 0
507 04/06/2010 07/03/2010 10/06/2010 3 0 0
508 02/03/2010 05/09/2010 08/11/2010 3 1 1
509 05/15/2010 08/15/2010 11/15/2010 3 2 0
510 01/12/2010 04/13/2010 07/14/2010 2 1 0
511 04/10/2010 07/10/2010 10/13/2010 3 1 1
512 04/01/2010 07/01/2010 10/01/2010 3 2 0
601 06/03/2010 09/09/2010 12/09/2010 3 1 0
602 02/28/2010 03/25/2010 06/23/2010 2 0 0
603 01/12/2010 04/12/2010 07/10/2010 2 0 0
604 03/13/2010 06/14/2010 09/15/2010 2 1 1
605 04/23/2010 07/24/2010 10/26/2010 3 3 0
606 02/22/2010 05/25/2010 08/23/2010 2 0 0
607 06/10/2010 09/10/2010 12/09/2010 3 1 0
608 02/20/2010 05/20/2010 08/17/2010 3 1 0
609 05/22/2010 08/25/2010 11/23/2010 3 3 2
610 01/09/2010 04/10/2010 07/12/2010 2 0 0
611 06/19/2010 09/21/2010 12/20/2010 3 1 1
612 03/12/2010 06/15/2010 09/16/2010 3 1 1
701 05/12/2010 08/10/2010 11/10/2010 2 0 0
702 02/09/2010 05/10/2010 08/12/2010 3 0 0
703 05/19/2010 08/20/2010 11/20/2010 3 1 0
704 06/01/2010 09/09/2010 12/10/2010 2 3 0
705 05/20/2010 08/20/2010 11/19/2010 3 1 1
706 06/12/2010 09/11/2010 12/10/2010 3 1 2
707 01/12/2010 04/12/2010 07/10/2010 2 1 1
708 03/13/2010 06/14/2010 09/15/2010 3 1 1
709 04/23/2010 07/24/2010 10/23/2010 3 1 0
710 03/16/2010 06/16/2010 09/16/2010 2 2 2
711 04/22/2010 07/22/2010 09/20/2010 2 0 0
712 06/10/2010 09/09/2010 12/09/2010 2 1 0
;
run;
```

Appendix B: SDTM Metadata

In this appendix, you can see the SDTM metadata that we used in Chapters 2 and 3. There are nine tabs in the SDTM metadata spreadsheet. Those tabs are shown here. If you go to the author page for this book, you will find this metadata stored in SDTM_METADATA.xlsx. See support.sas.com/publishing/authors/index.html.

Appendix B.1 - Define Header Metadata

	FILEOID	STUDYOID	STUDYNAME	STUDYDESCRIPTION	PROTOCOLNAME	STANDARD	VERSION	STYLESHEET
1	XYZ123	123	XYZ123	A PHASE IIB, DOUBLE-BLIND, MULTI-CENTER, PLACEBO CONTROLLED, PARALLEL GROUP TRIAL OF ANALGEZIA HCL FOR THE TREATMENT OF CHRONIC PAIN	XYZ123	SDTM-IG	3.2	define2-0-0.xsl
2								

Appendix B.2 - Table of Contents Metadata

	OID	NAME	REPEATING	ISREFERENCEDATA	PURPOSE	LABEL	STRUCTURE	CLASS	ARCHIVELOCATIONID	COMMENTOID
1	AE	AE	Yes	No	Tabulation	Adverse Events	Events - One record per event per subject	Events	ae	com.ae
2	DM	DM	No	No	Tabulation	Demographics	Special Purpose - One record per event per subject	Special Purpose	dm	
3	EX	EX	Yes	No	Tabulation	Exposure	One record per constant dosing interval per subject	Interventions	ex	
4	LB	LB	Yes	No	Tabulation	Laboratory Test Results	Findings - One record per lab test per subject	Findings	lb	
5	XP	XP	Yes	No	Tabulation	Pain Scores	One record per subject per visit	Findings	xp	
6	TA	TA	Yes	Yes	Tabulation	Trial Arms	One record per planned Element per Arm	Trial Design	ta	
7	TD	TD	Yes	Yes	Tabulation	Trial Disease Assessments	One record per planned schedule of assessments	Trial Design	td	
8	TE	TE	Yes	Yes	Tabulation	Trial Elements	One record per planned Element	Trial Design	te	
9	TI	TI	Yes	Yes	Tabulation	Trial Inclusion/Exclusion Criteria	One record per I/E criterion	Trial Design	ti	
10	TS	TS	Yes	Yes	Tabulation	Trial Summary	One record per trial summary parameter value	Trial Design	ts	
11	TV	TV	Yes	Yes	Tabulation	Trial Visits	One record per planned Visit per Arm	Trial Design	tv	
12	SUPPDM	SUPPDM	Yes	No	Tabulation	DM - Supplemental Qualifiers	Supplemental Qualifier - One record per qualifier	Relationship	suppdm	

Appendix B.3 - Variable-Level Metadata

	DOMAIN	VARNUM	VARIABLE	TYPE	LENGTH	LABEL	KEYSEQUENCE	SIGNIFICANT DIGITS	ORIGIN	COMMENTOID	DISPLAY FORMAT	COMPUTATION METHODOID	CODELIST NAME	MANDATORY	ROLE
1															
2	AE	1	STUDYID	text	15	Study Identifier	1		Assigned					Yes	Identifier
3	AE	2	DOMAIN	text	2	Domain Abbreviation			Derived			SETDOMAIN		Yes	Identifier
4	AE	3	USUBJID	text	25	Unique Subject Identifier	2		Assigned					Yes	Identifier
5	AE	4	AESEQ	integer	8	Sequence Number			Derived			AE.AESEQ		Yes	Identifier
6	AE	5	AETERM	text	200	Reported Term for the Adverse Event			CRF page 6					Yes	Topic
7	AE	6	AELLT	text	200	Lowest Level Term			Assigned					Yes	Synonym Qualifier
8	AE	7	AELLTCD	integer	8	Lowest Level Term Code			Assigned					Yes	Synonym Qualifier
9	AE	8	AEDECOD	text	200	Dictionary-Derived Term	3		Assigned				AEDECOD	Yes	Synonym Qualifier
10	AE	9	AEPTCD	integer	8	Preferred Term Code			Assigned					Yes	Variable Qualifier
11	AE	10	AEHLT	text	200	High Level Term			Assigned					Yes	Variable Qualifier
12	AE	11	AEHLTCD	integer	8	High Level Term Code			Assigned					Yes	Variable Qualifier
13	AE	12	AEHLGT	text	200	High Level Group Term			Assigned					Yes	Variable Qualifier
14	AE	13	AEHLGTCD	integer	8	High Level Group Term Code			Assigned					Yes	Variable Qualifier
15	AE	14	AEBODSYS	text	200	Body System or Organ Class			Assigned				AEBODSYS	No	Record Qualifier
16	AE	15	AEBDSYCD	integer	8	Body System or Organ Class Code			Assigned					Yes	Record Qualifier
17	AE	16	AESOC	text	200	Primary System Organ Class			Assigned					Yes	Record Qualifier
18	AE	17	AESOCCD	integer	8	Primary System Organ Class Code			Assigned					Yes	Record Qualifier
19	AE	18	AESEV	text	20	Severity			CRF page 6				AESEV	No	Record Qualifier
20	AE	19	AESER	text	2	Serious Event			CRF page 6				NY	No	Record Qualifier
21	AE	20	AEACN	text	40	Action Taken with Study Treatment			CRF page 6				ACN	No	Record Qualifier
22	AE	21	AEREL	text	40	Causality			CRF page 6				AEREL	No	Record Qualifier
23	AE	22	AESLIFE	text	2	Is Life Threatening			CRF page 6				NY	No	Record Qualifier
24	AE	23	EPOCH	text	40	Epoch			Derived			AE.EPOCH	EPOCH	No	Timing
25	AE	24	AESTDTC	date	16	Start Date/Time of Adverse Event	4		CRF page 6					No	Timing
26	AE	25	AEENDTC	date	16	End Date/Time of Adverse Event			CRF page 6					No	Timing
27	AE	26	AESTDY	integer	8	Study Day of Start of Adverse Event			Derived			STUDYDAYS		No	Timing
28	AE	27	AEENDY	integer	8	Study Day of End of Adverse Event			Derived			STUDYDAYS		No	Timing
29	DM	1	STUDYID	text	15	Study Identifier	1		Assigned					Yes	Identifier
30	DM	2	DOMAIN	text	2	Domain Abbreviation			Derived			SETDOMAIN		Yes	Identifier
31	DM	3	USUBJID	text	25	Unique Subject Identifier	2		Assigned					Yes	Identifier
32	DM	4	SUBJID	text	7	Subject Identifier for the Study			Assigned					Yes	Identifier

	DOMAIN	VARNUM	VARIABLE	TYPE	LENGTH	LABEL	KEYSEQUENCE	SIGNIFICANT DIGITS	ORIGIN	COMMENTOID	DISPLAY FORMAT	COMPUTATION METHODOID	CODELIST NAME	MANDATORY	ROLE
33	DM	5	RFSTDTC	date	16	Subject Reference Start Date/Time			CRF page 1					No	Record Qualifier
34	DM	6	RFENDTC	date	16	Subject Reference End Date/Time			Derived			DM.RFENDTC		No	Record Qualifier
35	DM	7	RFXSTDTC	date	16	Date/Time of First Study Treatment			CRF page 4					Yes	Record Qualifier
36	DM	8	RFXENDTC	date	16	Date/Time of Last Study Treatment			CRF page 4					Yes	Record Qualifier
37	DM	9	RFICDTC	date	16	Date/Time of Informed Consent			CRF Page 1					Yes	Record Qualifier
38	DM	10	RFPENDTC	date	16	Date/Time of End of Participation			CRF Page 30					Yes	Record Qualifier
39	DM	11	DTHDTC	date	16	Date/Time of Death			Derived			DM.DTHDTC		Yes	Record Qualifier
40	DM	12	DTHFL	text	2	Subject Death Flag			Derived			DM.DTHFL		Yes	Record Qualifier
41	DM	13	SITEID	text	7	Study Site Identifier			Assigned					Yes	Record Qualifier
42	DM	14	BRTHDTC	date	16	Date/Time of Birth			CRF page 1					No	Record Qualifier
43	DM	15	AGE	integer	8	Age			Derived		3.0	DM.AGE		No	Record Qualifier
44	DM	16	AGEU	text	10	Age Units			Derived			DM.AGEU	AGEU	No	Variable Qualifier
45	DM	17	SEX	text	2	Sex			CRF page 1				SEX	Yes	Record Qualifier
46	DM	18	RACE	text	80	Race			CRF page 1				RACE	No	Record Qualifier
47	DM	19	ARMCD	text	8	Planned Arm Code			eDT	COM.ARMCD			ARMCD	Yes	Record Qualifier
48	DM	20	ARM	text	40	Description of Planned Arm			eDT				ARM	Yes	Synonym Qualifier
49	DM	21	ACTARMCD	text	8	Actual Arm Code			eDT				ARMCD	Yes	Record Qualifier
50	DM	22	ACTARM	text	40	Description of Actual Arm			eDT				ARM	Yes	Synonym Qualifier
51	DM	23	COUNTRY	text	3	Country			Derived			DM.COUNTRY	COUNTRY	Yes	Record Qualifier
52	EX	1	STUDYID	text	15	Study Identifier	1		Assigned					Yes	Identifier
53	EX	2	DOMAIN	text	2	Domain Abbreviation			Derived			SETDOMAIN		Yes	Identifier
54	EX	3	USUBJID	text	25	Unique Subject Identifier	2		Assigned					Yes	Identifier
55	EX	4	EXSEQ	integer	8	Sequence Number			Derived			EX.EXSEQ		Yes	Identifier
56	EX	5	EXTRT	text	200	Name of Actual Treatment	3		Derived			EX.EXTRT		Yes	Topic
57	EX	6	EXDOSE	integer	8	Dose			CRF Page 4					No	Record Qualifier
58	EX	7	EXDOSU	text	40	Dose Units			Derived			EX.EXDOSU	UNIT	No	Variable Qualifier
59	EX	8	EXDOSFRM	text	80	Dose Form			Derived			EX.EXDOSFRM	FRM	No	Record Qualifier
60	EX	9	EPOCH	text	40	Epoch			Derived			EX.EPOCH	EPOCH	No	Timing
61	EX	10	EXSTDTC	date	16	Start Date/Time of Treatment			CRF Page 4					No	Timing
62	EX	11	EXENDTC	date	16	End Date/Time of Treatment	4		CRF Page 4					No	Timing
63	EX	12	EXSTDY	integer	8	Study Day of Start of Treatment			Derived			STUDYDAYS		No	Timing
64	EX	13	EXENDY	integer	8	Study Day of End of Treatment			Derived			STUDYDAYS		No	Timing

	DOMAIN	VARNUM	VARIABLE	TYPE	LENGTH	LABEL	KEYSEQUENCE	SIGNIFICANT DIGITS	ORIGIN	COMMENTOID	DISPLAY FORMAT	COMPUTATION METHODOID	CODELIST NAME	MANDATORY	ROLE
65	SUPPDM	1	STUDYID	text	15	Study Identifier	1		Assigned					Yes	
66	SUPPDM	2	RDOMAIN	text	2	Related Domain Abbreviation	2		Derived			SETDOMAIN		Yes	
67	SUPPDM	3	USUBJID	text	25	Unique Subject Identifier	3		Assigned					Yes	
68	SUPPDM	4	IDVAR	text	8	Identifying Variable	4		Assigned					No	
69	SUPPDM	5	IDVARVAL	text	200	Identifying Variable Value	5		Assigned					No	
70	SUPPDM	6	QNAM	text	8	Qualifier Variable Name	6		Assigned					Yes	
71	SUPPDM	7	QLABEL	text	40	Qualifier Variable Label			Assigned					Yes	
72	SUPPDM	8	QVAL	text	200	Data Value								Yes	
73	SUPPDM	9	QORIG	text	20	Origin			Derived			SETNULL		Yes	
74	SUPPDM	10	QEVAL	text	8	Evaluator			Derived			SETNULL		No	
75	LB	1	STUDYID	text	15	Study Identifier	1		Assigned					Yes	Identifier
76	LB	2	DOMAIN	text	2	Domain Abbreviation			Derived			SETDOMAIN		Yes	Identifier
77	LB	3	USUBJID	text	25	Unique Subject Identifier	2		Assigned					Yes	Identifier
78	LB	4	LBSEQ	integer	8	Sequence Number			Derived			LB.LBSEQ		Yes	Identifier
79	LB	5	LBTESTCD	text	8	Lab Test or Examination Short Name	4		Derived			LB.LBTESTCD	LBTESTCD	Yes	Topic
80	LB	6	LBTEST	text	40	Lab Test or Examination Name			Derived			LB.LBTEST	LBTEST	Yes	Synonym Qualifier
81	LB	7	LBCAT	text	40	Category for Lab Test	3		Derived			LB.LBCAT	LBCAT	No	Grouping Qualifier
82	LB	8	LBORRES	text	200	Result or Finding in Original Units			eDT					No	Result Qualifier
83	LB	9	LBORRESU	text	40	Original Units			eDT				UNIT	No	Variable Qualifier
84	LB	10	LBORNRLO	text	40	Reference Range Lower Limit in Orig Unit			eDT					No	Variable Qualifier
85	LB	11	LBORNRHI	text	40	Reference Range Upper Limit in			eDT					No	Variable Qualifier
86	LB	12	LBSTRESC	text	200	Character Result/Finding in Std			Derived			LB.LBSTRESC		No	Result Qualifier
87	LB	13	LBSTRESN	float	8	Numeric Result/Finding in		2	Derived			LB.LBSTRESN		No	Result Qualifier
88	LB	14	LBSTRESU	text	40	Standard Units			Derived			LB.LBSTRESU	UNIT	No	Variable Qualifier
89	LB	15	LBSTNRLO	float	8	Reference Range Lower Limit-Std		2	Derived			LB.LBSTNRLO		No	Variable Qualifier
90	LB	16	LBSTNRHI	float	8	Reference Range Upper Limit-Std		2	Derived			LB.LBSTNRHI		No	Variable Qualifier
91	LB	17	LBNRIND	text	20	Reference Range Indicator			Derived			LB.LBNRIND	LBNRIND	No	Variable Qualifier
92	LB	18	LBBLFL	text	2	Baseline Flag			Derived			LB.LBBLFL	NY	No	Record Qualifier
93	LB	19	VISITNUM	integer	8	Visit Number	5		eDT					No	Timing
94	LB	20	VISIT	text	40	Visit Name			eDT				VISIT	No	Timing
95	LB	21	EPOCH	text	40	Epoch			Derived			LB.EPOCH	EPOCH	No	Timing

	DOMAIN	VARNUM	VARIABLE	TYPE	LENGTH	LABEL	KEYSEQUENCE	SIGNIFICANT DIGITS	ORIGIN	COMMENTOID	DISPLAY FORMAT	COMPUTATION METHODOID	CODELIST NAME	MANDATORY	ROLE
96	LB	22	LBDTC	date	16	Date/Time of Specimen Collection			eDT					No	Timing
97	LB	23	LBDY	integer	8	Study Day of Specimen Collection			Derived			STUDYDAYS		No	Timing
98	TA	1	STUDYID	text	15	Study Identifier	1		Assigned					Yes	Identifier
99	TA	2	DOMAIN	text	2	Domain Abbreviation			Derived			SETDOMAIN		Yes	Identifier
100	TA	3	ARMCD	text	8	Planned Arm Code	2		Derived			METADATAENTI	ARMCD	Yes	Topic
101	TA	4	ARM	text	40	Description of Planned Arm			Derived			METADATAENTI	ARM	Yes	Synonym Qualifier
102	TA	5	TAETORD	integer	8	Planned Order of Element within	3		Derived			METADATAENTRY		Yes	Identifier
103	TA	6	ETCD	text	8	Element Code			Derived			METADATAENTRY		Yes	Record Qualifier
104	TA	7	ELEMENT	text	40	Description of Element			Derived			METADATAENTRY		No	Synonym Qualifier
105	TA	8	TABRANCH	text	200	Branch			Derived			METADATAENTRY		No	Rule
106	TA	9	TATRANS	text	200	Transition Rule			Derived			METADATAENTRY		No	Rule
107	TA	10	EPOCH	text	40	Epoch			Derived			METADATAENTI	EPOCH	No	Timing
108	TE	1	STUDYID	text	15	Study Identifier	1		Assigned					Yes	Identifier
109	TE	2	DOMAIN	text	2	Domain Abbreviation			Derived			SETDOMAIN		Yes	Identifier
110	TE	3	ETCD	text	8	Element Code	2		Derived			METADATAENTRY		Yes	Topic
111	TE	4	ELEMENT	text	40	Description of Element			Derived			METADATAENTRY		Yes	Synonym Qualifier
112	TE	5	TESTRL	text	200	Rule for Start of Element			Derived			METADATAENTRY		Yes	Rule
113	TE	6	TEENRL	text	200	Rule for End of Element			Derived			METADATAENTRY		No	Rule
114	TE	7	TEDUR	text	64	Planned Duration of Element			Derived			METADATAENTRY		No	Timing
115	TI	1	STUDYID	text	15	Study Identifier	1		Assigned					Yes	Identifier
116	TI	2	DOMAIN	text	2	Domain Abbreviation			Derived			SETDOMAIN		Yes	Identifier
117	TI	3	IETESTCD	text	8	Inclusion/Exclusion Criterion Short	2		Derived			METADATAENTRY		Yes	Topic
118	TI	4	IETEST	text	200	Inclusion/Exclusion Criterion			Derived			METADATAENTRY		Yes	Synonym Qualifier
119	TI	5	IECAT	text	40	Inclusion/Exclusion Category			Derived			METADATAENTI	IECAT	Yes	Grouping
120	TD	1	STUDYID	text	15	Study Identifier	1		Assigned					Yes	Identifier
121	TD	2	DOMAIN	text	2	Domain Abbreviation			Derived			SETDOMAIN		Yes	Identifier
122	TD	3	TDORDER	integer	8	Sequence of Planned Assessment Schedule	2		Derived			METADATAENTRY		Yes	Timing
123	TD	4	TDANCVAR	text	8	Anchor Variable Name			Derived			METADATAENTRY		Yes	Timing
124	TD	5	TDSTOFF	date	16	Offset from the Anchor			Derived			METADATAENTRY		Yes	Timing
125	TD	6	TDTGTPAI	date	16	Planned Assessment Interval			Derived			METADATAENTRY		Yes	Timing
126	TD	7	TDMINPAI	date	16	Planned Assessment Interval			Derived			METADATAENTRY		Yes	Timing

	DOMAIN	VARNUM	VARIABLE	TYPE	LENGTH	LABEL	KEYSEQUENCE	SIGNIFICANT DIGITS	ORIGIN	COMMENTOID	DISPLAY FORMAT	COMPUTATION METHODOID	CODELIST NAME	MANDATORY	ROLE
127	TD	8	TDMAXPAI	date	16	Planned Assessment Interval			Derived			METADATAENTRY		Yes	Timing
128	TD	9	TDNUMRPT	integer	8	Maximum Number of Actual Assessments			Derived			METADATAENTRY		Yes	Record Qualifier
129	TS	1	STUDYID	text	15	Study Identifier	1		Assigned					Yes	Identifier
130	TS	2	DOMAIN	text	2	Domain Abbreviation			Derived			SETDOMAIN		Yes	Identifier
131	TS	3	TSSEQ	integer	8	Sequence Number	2		Derived			METADATAENTRY		Yes	Identifier
132	TS	4	TSPARMCD	text	8	Trial Summary Parameter Short			Derived			METADATAENTI	TSPARMCD	Yes	Topic
133	TS	5	TSPARM	text	40	Trial Summary Parameter			Derived			METADATAENTI	TSPARM	Yes	Synonym Qualifier
134	TS	6	TSVAL	text	200	Parameter Value			Derived			METADATAENTRY		Yes	Result Qualifier
135	TS	8	TSVALCD	text	40	Parameter Value Code			Derived			METADATAENTRY		Yes	Result Qualifier
136	TS	9	TSVCDREF	text	40	Name of the Reference Terminology			Derived			METADATAENTRY		Yes	Result Qualifier
137	TS	10	TSVCDVER	text	40	Version of the Reference			Derived			METADATAENTRY		Yes	Result Qualifier
138	TV	1	STUDYID	text	15	Study Identifier	1		Assigned					Yes	Identifier
139	TV	2	DOMAIN	text	2	Domain Abbreviation			Derived			SETDOMAIN		Yes	Identifier
140	TV	3	VISITNUM	integer	8	Visit Number	2		Derived			METADATAENTRY		Yes	Topic
141	TV	4	VISIT	text	40	Visit Name			Derived			METADATAENTRY		No	Synonym Qualifier
142	TV	5	ARMCD	text	8	Planned Arm Code	3		Derived			METADATAENTI	ARMCD	No	Record Qualifier
143	TV	6	ARM	text	40	Description of Planned Arm			Derived			METADATAENTI	ARM	No	Synonym Qualifier
144	TV	7	TVSTRL	text	200	Visit Start Rule			Derived			METADATAENTRY		Yes	Rule
145	TV	8	TVENRL	text	200	Visit End Rule			Derived			METADATAENTRY		No	Rule
146	XP	1	STUDYID	text	15	Study Identifier	1		Assigned					Yes	Identifier
147	XP	2	DOMAIN	text	2	Domain Abbreviation			Derived			SETDOMAIN		Yes	Identifier
148	XP	3	USUBJID	text	25	Unique Subject Identifier	2		Assigned					Yes	Identifier
149	XP	4	XPSEQ	integer	8	Sequence Number			Derived			XP.XPSEQ		Yes	Identifier
150	XP	5	XPTESTCD	text	8	Pain Test Short Name	3		Derived			XP.XPTESTCD	XPTESTCD	Yes	Topic
151	XP	6	XPTEST	text	40	Pain Test Name			Derived			XP.XPTEST	XPTEST	Yes	Synonym Qualifier
152	XP	7	XPORRES	text	200	Result or Finding in Original Units			CRF Page 5					No	Result Qualifier
153	XP	8	XPSTRESC	text	200	Character Result/Finding in Std			CRF Page 5					No	Result Qualifier
154	XP	9	XPSTRESN	integer	8	Numeric Result/Finding in Std			CRF Page 5					No	Result Qualifier
155	XP	10	EPOCH	text	40	Epoch			Derived			XP.EPOCH	EPOCH	No	Timing
156	XP	11	VISITNUM	integer	8	Visit Number	4		Assigned					No	Timing
157	XP	12	VISIT	text	40	Visit Name			CRF Page 5				VISIT	No	Timing
158	XP	13	XPBLFL	text	2	Baseline Flag			Derived			XP.XPBLFL	NY	No	Record Qualifier
159	XP	14	XPDTC	date	16	Date/Time of Collection			CRF Page 5					No	Timing
160	XP	15	XPDY	integer	8	Study Day of Collection			Derived			STUDYDAYS		No	Timing

Appendix B.4 - Value-Level Metadata

	DOMAIN	VARIABLE	WHERECLAUSEOID	VALUEVAR	VARNUM	VALUENAME	TYPE	LENGTH	LABEL	SIGNIFICANT DIGITS	ORIGIN	COMMENT OID	DISPLAY FORMAT	COMPUTATIONMETHOD OID	CODELIST NAME	MANDATORY	ROLE
2	LB	LBORRES		LBTESTCD	1	ALB	float	8	Albumin, Microalbumin	2	eDT		4.2			No	
3	LB	LBORRES		LBTESTCD	2	ALP	float	8	Alkaline Phosphatase	0	eDT		3			No	
4	LB	LBORRES		LBTESTCD	3	ALT	float	8	Aminotransferase; SGPT	0	eDT		3			No	
5	LB	LBORRES		LBTESTCD	4	AST	float	8	Aspartate Aminotransferase, SGOT	0	eDT		3			No	
6	LB	LBORRES		LBTESTCD	5	BILDIR	float	8	Direct Bilirubin	2	eDT		4.2			No	
7	LB	LBORRES		LBTESTCD	6	BILI	float	8	Bilirubin; Total Bilirubin	2	eDT		4.2			No	
8	LB	LBORRES		LBTESTCD	7	GGT	float	8	Gamma Glutamyl Transferase	0	eDT		3			No	
9	LB	LBORRES		LBTESTCD	8	HCT	float	8	Hematocrit	0	eDT		3			No	
10	LB	LBORRES		LBTESTCD	9	HGB	float	8	Hemoglobin	2	eDT		5.2			No	
11	LB	LBORRES		LBTESTCD	10	PROT	float	8	Protein	2	eDT		4.2			No	
12	LB	LBORRES	WC.LB.LBTESTCD.GLUC.LBCAT.CHEMISTRY	LBTESTCD	11	GLUC	float	8	Glucose	2	eDT		4.2			No	
13	LB	LBORRES	WC.LB.LBTESTCD.GLUC.LBCAT.URINALYSIS	LBTESTCD	12	GLUC	text	8	Glucose		eDT				URINGLUC	No	
14	SUPPDM	QVAL		QNAM	1	RACEOTH	text	20	Race, Other		Derived			SUPPDM.RACEOTH		No	
15	SUPPDM	QVAL		QNAM	2	RANDDTC	text	10	Randomization Date		Derived			SUPPDM.RANDDT		No	

Appendix B.5 – Computational Method Metadata

	COMPUTATIONMETHODOID	LABEL	TYPE	COMPUTATIONMETHOD
2	SETDOMAIN	Hardcode to assign domain	Computation	Hardcode to the name of the SDTM domain
3	STUDYDAYS	--DY derivation	Computation	If the date is before RFSTDTC, then --DY = this date minus RFSTDTC. Otherwise, --DY = this date minus RFSTDTC plus 1.
4	SETNULL	Null Value	Computation	Set to null
5	AE.AESEQ	AESEQ derivation	Computation	Unique sequence based on sort order of studyid usubjid aedecod aestdtc aeendtc
6	AE.EPOCH	EPOCH hardcode	Computation	Hardcoded to TREATMENT
7	DM.RFENDTC	RFENDTC derivation	Computation	RFENDTC is the last date of dosing from the source dosing dataset
8	DM.DTHDTC	DTHDTC derivation	Computation	DTHDTC is set to null as no subjects died
9	DM.DTHFL	DTHFL derivation	Computation	DTHFL='N' since no subjects died
10	DM.AGE	Algorithm to compute age	Computation	integer value of (BRTHDTC - RFSTDTC)/365.25
11	DM.AGEU	Age units hardcode	Computation	Hardcoded to YEARS
12	DM.COUNTRY	Country hardcode	Computation	Hardcoded to USA
13	EX.EXSEQ	EXSEQ derivation	Computation	Unique sequence based on sort order of studyid usubjid extrt exstdtc
14	EX.EXTRT	EXTRT derivation	Computation	All subjects were dosed with what they were randomized to, so EXTRT=DM.ARM
15	EX.EXDOSU	EXDOSU hardcode	Computation	Hardcoded to mg
16	EX.EXDOSFRM	EXDOSFRM hardcode	Computation	Hardcoded to TABLET, COATED
17	EX.EPOCH	EPOCH hardcode	Computation	Hardcoded to TREATMENT
18	LB.LBSEQ	LBSEQ derivation	Computation	Unique sequence based on sort order of studyid usubjid lbcat lbtestcd visitnum
19	LB.LBTESTCD	LBTESTCD derivation	Computation	Source labs.labtest variable mapped to LBTESTCD controlled terminology
20	LB.LBTEST	LBTEST derivation	Computation	Source labs.labtest variable mapped to LBTEST controlled terminology
21	LB.LBCAT	LBCAT derivation	Computation	Source labs.labcat variable mapped to LBCAT controlled terminology
22	LB.LBSTRESC	LBSTRESC derivation	Computation	Copied directly (and unrealistically) from LBORRES
23	LB.LBSTRESN	LBSTRESN derivation	Computation	Copied directly (and unrealistically) from LBORRES
24	LB.LBSTRESU	LBSTRESN derivation	Computation	Copied directly (and unrealistically) from LBORRESU
25	LB.LBSTNRLO	LBSTNRLO derivation	Computation	Set equal to (and unrealistically) from the source labs.lownorm
26	LB.LBSTNRHI	LBSTNRHI derivation	Computation	Set equal to (and unrealistically) from the source labs.highnorm

	COMPUTATIONMETHODOID	LABEL	TYPE	COMPUTATIONMETHOD
27	LB.LBNRIND	LBNRIND derivation	Computation	if lbtestcd = 'GLUC' and lbcat = 'URINALYSIS' and lborres = 'POSITIVE' then lbnrind = 'HIGH'; else if lbtestcd = 'GLUC' and lbcat = 'URINALYSIS' and lborres = 'NEGATIVE' then lbnrind = 'NORMAL'; else if lbstnrlo ne . and lbstresn ne . and round(lbstresn,.0000001) < round(lbstnrlo,.0000001) then lbnrind = 'LOW'; else if lbstnrhi ne . and lbstresn ne . and round(lbstresn,.0000001) > round(lbstnrhi,.0000001) then lbnrind = 'HIGH'; else if lbstnrhi ne . and lbstresn ne . then lbnrind = 'NORMAL';
28	LB.LBBLFL	LBBLFL derivation	Computation	If visit=Baseline then Y, else blank
29	LB.EPOCH	EPOCH derivation	Computation	If VISITNUM < 0 then SCREENING and otherwise TREATMENT
30	METADATAENTRY	Manual entry	Computation	This variable consists of study level metadata that was manually entered
31	XP.XPSEQ	XPSEQ derivation	Computation	Unique sequence based on sort order of studyid usubjid xptestcd visitnum
32	XP.XPTESTCD	XPTESTCD hardcode	Computation	Set to XPAIN
33	XP.XPTEST	XPTEST hardcode	Computation	Set to Pain Score
34	XP.EPOCH	EPOCH hardcode	Computation	Hardcoded to TREATMENT
35	XP.XPBLFL	XPBLFL derivation	Computation	If visit=Baseline then Y, else blank
36	SUPPDM.RACEOTH	RACEOTH derivation	Computation	Set to other race from source demographics file
37	SUPPDM.RANDDT	RANDDT derivation	Computation	Set equal to the randomization date from the source demographics file

Appendix B.6 - Codelist Metadata

CODELISTNAME	RANK	CODEDVALUE	TRANSLATED	TYPE	CODELIST DICTIONARY	CODELIST VERSION	ORDERNUMBER	sourcedataset	sourcevariable	sourcevalue	sourcetype
ACN	1	DOSE INCREASED	DOSE INCREASED	text				adverse	aeaction	3	number
ACN	2	DOSE NOT CHANGED	DOSE NOT CHANGED	text				adverse	aeaction	4	number
ACN	3	DOSE REDUCED	DOSE REDUCED	text				adverse	aeaction	2	number
ACN	4	DRUG INTERRUPTED	DRUG INTERRUPTED	text				adverse	aeaction	1	number
ACN	5	DRUG WITHDRAWN	DRUG WITHDRAWN	text				adverse	aeaction		number
ACN	6	NOT APPLICABLE	NA	text				adverse	aeaction		number
ACN	7	UNKNOWN	U; Unknown	text				adverse	aeaction	5	number
AEBODSYS	1			text	MedDRA	18.0					
AEDECOD	1			text	MedDRA	18.0					
AEREL	1	NOT RELATED	NOT RELATED	text				adverse	aerel	1	number
AEREL	2	POSSIBLY RELATED	POSSIBLY RELATED	text				adverse	aerel	2	number
AEREL	3	PROBABLY RELATED	PROBABLY RELATED	text				adverse	aerel	3	number
AESEV	1	MILD	Grade 1; 1	text				adverse	aesev	1	number
AESEV	2	MODERATE	Grade 2; 2	text				adverse	aesev	2	number
AESEV	3	SEVERE	Grade 3; 3	text				adverse	aesev	3	number
AGEU	1	YEARS	YEARS	text						NONE	
ARM	1	Analgezia HCL 30 mg	Analgezia HCL 30 mg	text				demographic	trt	1	number
ARM	2	Placebo	Placebo	text				demographic	trt	0	number
ARMCD	1	ALG123	Analgezia HCL 30 mg	text				demographic	trt	1	number
ARMCD	2	PLACEBO	Placebo	text				demographic	trt	0	number
COUNTRY	1	USA	UNITED STATES	text						NONE	
FRM	1	TABLET, COATED	TABLET, COATED	text							
IECAT	1	INCLUSION	INCLUSION	text							
IECAT	2	EXCLUSION	EXCLUSION	text							
LBCAT	1	CHEMISTRY	CHEMISTRY	text				labs	labcat	CHEMISTRY	character
LBCAT	2	HEMATOLOGY	HEMATOLOGY	text				labs	labcat	HEMATOLOG	character
LBCAT	3	URINALYSIS	URINALYSIS	text				labs	labcat	URINALYSIS	character

CODELISTNAME	RANK	CODEDVALUE	TRANSLATED	TYPE	CODELIST DICTIONARY	CODELIST VERSION	ORDERNUMBER	sourcedataset	sourcevariable	sourcevalue	sourcetype
LBNRIND	1	HIGH	HIGH	text							
LBNRIND	2	LOW	LOW	text							
LBNRIND	3	NORMAL	NORMAL	text							
LBTEST	1	Albumin; Microalbumin	Albumin; Microalbumin	text				labs	labtest	ALBUMIN	character
LBTEST	2	Alkaline Phosphatase	Alkaline Phosphatase	text				labs	labtest	ALK. PHOS.	character
LBTEST	3	Alanine Aminotransferase, SGPT	Alanine Aminotransferase, SGPT	text				labs	labtest	ALT (SGPT)	character
LBTEST	4	Aspartate Aminotransferase; SGOT	Aspartate Aminotransferase; SGOT	text				labs	labtest	AST (SGOT)	character
LBTEST	5	Direct Bilirubin	Direct Bilirubin	text				labs	labtest	DIRECT BILI	character
LBTEST	6	Bilirubin; Total Bilirubin	Bilirubin; Total Bilirubin	text				labs	labtest	TOTAL BILI	character
LBTEST	7	Gamma Glutamyl Transferase	Gamma Glutamyl Transferase	text				labs	labtest	GGTP	character
LBTEST	8	EVF; Erythrocyte Volume Fraction; Hematocrit; PCV; Packed Cell Volume	EVF; Erythrocyte Volume Fraction; Hematocrit; PCV; Packed Cell Volume	text				labs	labtest	HEMATOCRIT	character
LBTEST	9	FHGB; Free Hemoglobin; Hemoglobin	FHGB; Free Hemoglobin; Hemoglobin	text				labs	labtest	HEMOGLOBII	character
LBTEST	10	Protein	Protein	text				labs	labtest	TOTAL PROT	character
LBTEST	11	Glucose	Glucose	text				labs	labtest	GLUCOSE	character
LBTESTCD	1	ALB	Albumin; Microalbumin	text				labs	labtest	ALBUMIN	character
LBTESTCD	2	ALP	Alkaline Phosphatase	text				labs	labtest	ALK. PHOS	character
LBTESTCD	3	ALT	Alanine Aminotransferase, SGPT	text				labs	labtest	ALT (SGPT)	character
LBTESTCD	4	AST	Aspartate Aminotransferase; SGOT	text				labs	labtest	AST (SGOT)	character
LBTESTCD	5	BILDIR	Direct Bilirubin	text				labs	labtest	DIRECT BILI	character
LBTESTCD	6	BILI	Bilirubin; Total Bilirubin	text				labs	labtest	TOTAL BILI	character
LBTESTCD	7	GGT	Gamma Glutamyl Transferase	text				labs	labtest	GGTP	character
LBTESTCD	8	HCT	EVF; Erythrocyte Volume Fraction; Hematocrit; PCV; Packed Cell	text				labs	labtest	HEMATOCRIT	character
LBTESTCD	9	HGB	FHGB; Free Hemoglobin; Hemoglobin	text				labs	labtest	HEMOGLOBII	character
LBTESTCD	10	PROT	Protein	text				labs	labtest	TOTAL PROT	character

1	CODELISTNAME	RANK	CODEDVALUE	TRANSLATED	TYPE	CODELIST DICTIONARY	CODELIST VERSION	ORDERNUMBER	sourcedataset	sourcevariable	sourcevalue	sourcetype
53	LBTESTCD	11	GLUC	Glucose	text				labs	labtest	GLUCOSE	character
54	NY	1	N	No	text				adverse	serious	NO	character
55	NY	2	NA	NA	text				adverse	serious		character
56	NY	3	U	U: Unknown	text				adverse	serious		character
57	NY	4	Y	Yes	text				adverse	serious		character
58	RACE	1	BLACK OR AFRICAN AMERICAN	BLACK OR AFRICAN AMERICAN	text				demographic	race	2	number
59	RACE	2	ASIAN	ASIAN	text				demographic	race	3	number
60	RACE	3	WHITE	WHITE	text				demographic	race	1	number
61	SEX	1	F	FEMALE	text				demographic	gender	2	number
62	SEX	2	M	MALE	text				demographic	gender	1	number
63	SEX	3	U	UNKNOWN	text				demographic	gender		number
64	TSPARM	1	Added on to Existing Treatments	Added on to Existing Treatments	text							
65	TSPARM	2	Planned Maximum Age of Subjects	Planned Maximum Age of Subjects	text							
66	TSPARM	3	Planned Minimum Age of Subjects	Planned Minimum Age of Subjects	text							
67	TSPARM	4	Planned Trial Length	Planned Trial Length	text							
68	TSPARM	5	Planned Number of Subjects	Planned Number of Subjects	text							
69	TSPARM	6	Trial is Randomized	Trial is Randomized	text							
70	TSPARM	7	Sex of Participants	Sex of Participants	text							
71	TSPARM	8	Study Stop Rules	Study Stop Rules	text							
72	TSPARM	9	Trial Blinding Schema	Trial Blinding Schema	text							
73	TSPARM	10	Control Type	Control Type	text							
74	TSPARM	11	Diagnosis Group	Diagnosis Group	text							
75	TSPARM	12	Trial Indication Type	Trial Indication Type	text							
76	TSPARM	13	Trial Title	Trial Title	text							
77	TSPARM	14	Trial Phase Classification	Trial Phase Classification	text							

1	CODELISTNAME	RANK	CODEDVALUE	TRANSLATED	TYPE	CODELIST DICTIONARY	CODELIST VERSION	ORDERNUMBER	sourcedataset	sourcevariable	sourcevalue	sourcetype
102	TSPARM	39	Healthy Subject	Healthy Subject	text							
103	TSPARM	40	Stable Disease Minimum Duration	Stable Disease Minimum Duration	text							
104	TSPARM	41	Confirmed Response Minimum Duration	Confirmed Response Minimum Duration	text							
105	TSPARMCD	1	ADDON	Added on to Existing Treatments	text							
106	TSPARMCD	2	AGEMAX	Planned Maximum Age of Subjects	text							
107	TSPARMCD	3	AGEMIN	Planned Minimum Age of Subjects	text							
108	TSPARMCD	4	LENGTH	Planned Trial Length	text							
109	TSPARMCD	5	PLANSUB	Planned Number of Subjects	text							
110	TSPARMCD	6	RANDOM	Trial is Randomized	text							
111	TSPARMCD	7	SEXPOP	Sex of Participants	text							
112	TSPARMCD	8	STOPRULE	Study Stop Rules	text							
113	TSPARMCD	9	TBLIND	Trial Blinding Schema	text							
114	TSPARMCD	10	TCNTRL	Control Type	text							
115	TSPARMCD	11	TDIGRP	Diagnosis Group	text							
116	TSPARMCD	12	TINDTP	Trial Indication Type	text							
117	TSPARMCD	13	TITLE	Trial Title	text							
118	TSPARMCD	14	TPHASE	Trial Phase Classification	text							
119	TSPARMCD	15	TTYPE	Trial Type	text							
120	TSPARMCD	16	CURTRT	Current Therapy or Treatment	text							
121	TSPARMCD	17	OBJPRIM	Trial Primary Objective	text							
122	TSPARMCD	18	OBJSEC	Trial Secondary Objective	text							
123	TSPARMCD	19	SPONSOR	Clinical Study Sponsor	text							
124	TSPARMCD	20	INDIC	Trial Indication	text							
125	TSPARMCD	21	TRT	Investigational Therapy or Treatment	text							
126	TSPARMCD	22	RANDQT	Randomization Quotient	text							

	CODELISTNAME	RANK	CODEDVALUE	TRANSLATED	TYPE	CODELIST DICTIONARY	CODELIST VERSION	ORDERNUMBER	sourcedataset	sourcevariable	sourcevalue	sourcetype
78	TSPARM	15	Trial Type	Trial Type	text							
79	TSPARM	16	Current Therapy or Treatment	Current Therapy or Treatment	text							
80	TSPARM	17	Trial Primary Objective	Trial Primary Objective	text							
81	TSPARM	18	Trial Secondary Objective	Trial Secondary Objective	text							
82	TSPARM	19	Clinical Study Sponsor	Clinical Study Sponsor	text							
83	TSPARM	20	Trial Indication	Trial Indication	text							
84	TSPARM	21	Investigational Therapy or Treatment	Investigational Therapy or Treatment	text							
85	TSPARM	22	Randomization Quotient	Randomization Quotient	text							
86	TSPARM	23	Registry Identifier	Registry Identifier	text							
87	TSPARM	24	Primary Outcome Measure	Primary Outcome Measure	text							
88	TSPARM	25	Secondary Outcome Measure	Secondary Outcome Measure	text							
89	TSPARM	26	Exploratory Outcome Measure	Exploratory Outcome Measure	text							
90	TSPARM	27	Pharmacological Class of Investigational Therapy	Pharmacological Class of Investigational Therapy	text							
91	TSPARM	28	Planned Country of Investigational Site(s)	Planned Country of Investigational Site(s)	text							
92	TSPARM	29	Adaptive Design	Adaptive Design	text							
93	TSPARM	30	Data Cutoff Date	Data Cutoff Date	text							
94	TSPARM	31	Data Cutoff Description	Data Cutoff Description	text							
95	TSPARM	32	Intervention Model	Intervention Model	text							
96	TSPARM	33	Planned Number of Arms	Planned Number of Arms	text							
97	TSPARM	34	Study Type	Study Type	text							
98	TSPARM	35	Intervention Type	Intervention Type	text							
99	TSPARM	36	Study Start Date	Study Start Date	text							
100	TSPARM	37	Study End Date	Study End Date	text							
101	TSPARM	38	Actual Number of Subjects	Actual Number of Subjects	text							

	CODELISTNAME	RANK	CODEDVALUE	TRANSLATED	TYPE	CODELIST DICTIONARY	CODELIST VERSION	ORDERNUMBER	sourcedataset	sourcevariable	sourcevalue	sourcetype
155	URINGLUC	2	POSITIVE	Positive	text				labs	labtest	1	number
156	EPOCH	1	SCREENING	SCREENING	text							
157	EPOCH	2	TREATMENT	TREATMENT	text							
158	XPTEST	1	Pain Score	Pain Score	text							
159	XPTESTCD	1	XPPAIN	XPPAIN	text							

	CODELISTNAME	RANK	CODEDVALUE	TRANSLATED	TYPE	CODELIST DICTIONARY	CODELIST VERSION	ORDERNUMBER	sourcedataset	sourcevariable	sourcevalue	sourcetype
127	TSPARMCD	23	REGID	Registry Identifier	text							
128	TSPARMCD	24	OUTMSPRI	Primary Outcome Measure	text							
129	TSPARMCD	25	OUTMSSEC	Secondary Outcome Measure	text							
130	TSPARMCD	26	OUTMSEXP	Exploratory Outcome Measure	text							
131	TSPARMCD	27	PCLAS	Pharmacological Class of Investigational Therapy	text							
132	TSPARMCD	28	FCNTRY	Planned Country of Investigational Site(s)	text							
133	TSPARMCD	29	ADAPT	Adaptive Design	text							
134	TSPARMCD	30	DCUTDTC	Data Cutoff Date	text							
135	TSPARMCD	31	DCUTDESC	Data Cutoff Description	text							
136	TSPARMCD	32	INTMODEL	Intervention Model	text							
137	TSPARMCD	33	NARMS	Planned Number of Arms	text							
138	TSPARMCD	34	STYPE	Study Type	text							
139	TSPARMCD	35	INTTYPE	Intervention Type	text							
140	TSPARMCD	36	SSTDTC	Study Start Date	text							
141	TSPARMCD	37	SENDTC	Study End Date	text							
142	TSPARMCD	38	ACTSUB	Actual Number of Subjects	text							
143	TSPARMCD	39	HLTSUBJI	Healthy Subject	text							
144	TSPARMCD	40	SDMDUR	Stable Disease Minimum Duration	text							
145	TSPARMCD	41	CRMDUR	Confirmed Response Minimum Duration	text							
146	UNIT	1	%	Percentage	text				labs	colunits	%	character
147	UNIT	2	g/dL	Gram per Deciliter	text				labs	colunits	g/dL	character
148	UNIT	3	IU/L	International Unit per Liter	text				labs	colunits	IU/L	character
149	UNIT	4	mg/dL	Milligram per Deciliter	text				labs	colunits	mg/dL	character
150	UNIT	5	mg	Milligram	text							
151	VISIT	1	Baseline	Baseline	text				labs	month	0	number
152	VISIT	2	Month 3	Month 3	text				labs	month	1	number
153	VISIT	3	Month 6	Month 6	text				labs	month	2	number
154	URINGLUC	1	NEGATIVE	Negative	text				labs	labtest	0	number

Appendix B.7 – Where Clause Metadata

	WHERECLAUSEOID	SEQ	SOFTHARD	ITEMOID	COMPARATOR	VALUES	COMMENTOID
2	WC.LB.LBTESTCD.GLUC.LBCAT.CHEMISTRY	1	Soft	LB.LBCAT	EQ	CHEMISTRY	
3	WC.LB.LBTESTCD.GLUC.LBCAT.CHEMISTRY	2	Soft	LB.LBTESTCD	EQ	GLUC	
4	WC.LB.LBTESTCD.GLUC.LBCAT.URINALYSIS	1	Soft	LB.LBCAT	EQ	URINALYSIS	
5	WC.LB.LBTESTCD.GLUC.LBCAT.URINALYSIS	2	Soft	LB.LBTESTCD	EQ	GLUC	

Appendix B.8 - Comment Metadata

	COMMENTOID	COMMENT
2	COM.ARMCD	Assigned based on Randomization Number.
3	com.ae	MedDRA dictionary version XX.1 was used to code all adverse events.

Appendix B.9 – External Links Metadata

	LeafID	LeafRelPath	LeafPageRef	LeafPageRefType	Title	SupplementalDoc	AnnotatedCRF
2	blankcrf	blankcrf.pdf			Annotated Case Report Form		Y
3	CRTRG	reviewersguide.pdf			CRT Reviewer's Guide	Y	

Appendix C: ADaM Metadata

In this appendix, you can see the ADaM metadata that we used in Chapter 6. There are ten tabs in the ADaM metadata spreadsheet. Those tabs are shown here. If you go to the author page for this book, you will find this metadata stored in ADAM_METADATA.xlsx. See support.sas.com/publishing/authors/index.html.

Appendix C.1 - Define Header Metadata

	A	B	C	D	E	F	G	H	I
	FILEOID	STUDYOID	STUDYNAME	STUDYDESCRIPTION	PROTOCOLNAME	STANDARD	VERSION	SCHEMALOCATION	STYLESHEET
	XYZ123	123	XYZ123	A PHASE IIB, DOUBLE-BLIND, MULTI-CENTER, PLACEBO CONTROLLED, PARALLEL GROUP TRIAL OF ANALGEZIA HCL FOR THE TREATMENT OF CHRONIC PAIN	XYZ123	ADaM-IG	1.0	http://www.cdisc.org/n s/odm/v1.2 util/adamres-draft2.xsd	define2-0-0.xsl

Appendix C.2 - Table of Contents Metadata

	REPEATING	ISREFERENCEDATA	PURPOSE	LABEL	STRUCTURE	DOMAINKEYS	CLASS	ARCHIVELOCATIONID	COMMENTOID
	Yes	No	Analysis	Adverse Events Analysis Datasets	ADAE - One record per event per subject	STUDYID, USUBJID, AEDECOD, ASTDT	Occurrence data structure	ADAE	COM.ADAE
	No	No	Analysis	Subject Level Analysis Dataset	ADSL - One record per subject	STUDYID, USUBJID	Subject level analysis dataset	ADSL	COM.ADSL
	Yes	No	Analysis	Efficacy/Pain Scores Analysis Dataset	One record per subject per visit	STUDYID, USUBJID, PARAMCD, AVISITN	Basic data structure	ADEF	COM.ADEF
	No	No	Analysis	Time-to-Pain Relief Analysis Dataset	One record per subject	STUDYID, USUBJID	Basic data structure	ADTTE	COM.ADTTE

Appendix C.3 - Variable-Level Metadata

	DOMAIN	VARNUM	VARIABLE	TYPE	LENGTH	LABEL	KEYSEQUENCE	SIGNIFICANTDIGITS	ORIGIN
2	ADSL	1	STUDYID	text	15	Study Identifier	1		DM
3	ADSL	2	USUBJID	text	25	Unique Subject Identifier	2		DM
4	ADSL	3	SUBJID	text	7	Subject Identifier for the Study			DM
5	ADSL	4	SITEID	text	7	Study Site Identifier			DM
6	ADSL	5	COUNTRY	text	3	Country			DM
7	ADSL	6	BRTHDT	integer	8	Date of Birth			DM.BIRTHDTC
8	ADSL	7	AGE	integer	8	Age			DM
9	ADSL	8	AGEU	text	5	Age Units			DM
10	ADSL	9	AGEGR1	text	40	Pooled Age Group 1			Derived
11	ADSL	10	AGEGR1N	integer	8	Pooled Age Group 1 (N)			Derived
12	ADSL	11	SEX	text	1	Sex			DM
13	ADSL	12	RACE	text	40	Race			DM
14	ADSL	13	RACEOTH	text	40	Race, Other, Specify			DM
15	ADSL	14	RANDDT	integer	8	Date of Randomization			Derived
16	ADSL	15	TRTSDT	integer	8	Date of First Exposure to Treatment			Derived
17	ADSL	16	TRTEDT	integer	8	Date of Last Exposure to Treatment			Derived
18	ADSL	17	ARM	text	40	Description of Planned Arm			DM
19	ADSL	18	TRT01P	text	40	Planned Treatment for Period 01			DM.ARM
20	ADSL	19	TRT01A	text	40	Actual Treatment for Period 01			DM.ARM
21	ADSL	20	TRT01PN	integer	8	Planned Treatment for Period 01 (N)			Derived
22	ADSL	21	TRT01AN	integer	8	Actual Treatment for Period 01 (N)			Derived
23	ADSL	22	ITTFL	text	1	Intent-To-Treat Population Flag			Derived
24	ADSL	23	SAFFL	text	1	Safety Population Flag			Derived
25	ADSL	24	RESPFL	text	1	Efficacy Responder Flag			Derived
26	ADEF	1	STUDYID	text	15	Study Identifier	1		ADSL
27	ADEF	2	USUBJID	text	25	Unique Subject Identifier	2		ADSL
28	ADEF	3	SITEID	text	7	Study Site Identifier			ADSL
29	ADEF	4	COUNTRY	text	3	Country			ADSL
30	ADEF	5	RANDDT	integer	8	Date of Randomization			ADSL
31	ADEF	6	AGE	integer	8	Age			ADSL
32	ADEF	7	AGEGR1N	integer	8	Pooled Age Group 1 (N)			ADSL

	COMMENTOID	DISPLAYFORMAT	COMPUTATIONMETHOD	CODELISTNAME	MANDATORY	ROLE	ROLECODELIST
2					Yes		
3					Yes		
4					Yes		
5					Yes		
6				COUNTRY	Yes		
7					No		
8		3.0			Yes		
9				AGEU	Yes		
10			AGEGR1	AGEGR1	No		
11			COMP.SEECODELIST	AGEGR1N	No		
12				SEX	Yes		
13				RACE	Yes		
14					No		
15			RANDDT		No		
16			TRTSDT		Yes		
17			TRTEDT		Yes		
18				ARM	Yes		
19				ARM	Yes		
20				ARM	Yes		
21			COMP.SEECODELIST	ARMN	No		
22			COMP.SEECODELIST	ARMN	No		
23			ITTFL	FLAGS	No		
24			SAFFL	FLAGS	No		
25			RESPFL	YN	No		
26					Yes		
27					Yes		
28					Yes		
29				COUNTRY	Yes		
30		date9.			No		
31					Yes		
32			COMP.SEECODELIST	AGEGR1N	Yes		
33			AGEGR1	AGEGR1	Yes		

	DOMAIN	VARNUM	VARIABLE	TYPE	LENGTH	LABEL	KEYSEQUENCE	SIGNIFICANTDIGITS	ORIGIN
33	ADEF	8	AGEGR1	text	20	Pooled Age Group 1			ADSL
34	ADEF	9	SEX	text	1	Sex			ADSL
35	ADEF	10	RACE	text	40	Race			ADSL
36	ADEF	11	TRTPN	integer	8	Planned Treatment (N)			ADSL.TRT01PN
37	ADEF	13	TRTP	text	40	Planned Treatment			ADSL.TRT01P
38	ADEF	14	PARAMCD	text	8	Parameter Code	3		Assigned
39	ADEF	15	PARAM	text	40	Parameter			Assigned
40	ADEF	16	AVISIT	text	16	Analysis Visit			XP.VISIT
41	ADEF	17	AVISITN	integer	8	Analysis Visit (N)	4		Derived
42	ADEF	18	ABLFL	text	1	Baseline Record Flag			Derived
43	ADEF	19	XPSEQ	integer	8	Sequence Number			XP
44	ADEF	20	VISITNUM	integer	8	Visit Number			XP
45	ADEF	21	ADT	integer	8	Analysis Date	5		Derived
46	ADEF	22	ADY	integer	8	Analysis Relative Day			Derived
47	ADEF	23	AVAL	integer	8	Analysis Value			Derived
48	ADEF	24	AVALC	text	12	Analysis Value (C)			Derived
49	ADEF	25	BASE	integer	8	Baseline Value			Derived
50	ADEF	27	CHG	integer	8	Change from Baseline			Derived
51	ADEF	28	CRIT1FL	text	1	Criterion 1 Evaluation Result Flag			Derived
52	ADEF	29	CRIT1	text	60	Analysis Criterion 1			Derived
53	ADEF	30	ITTFL	text	1	Intent-to-Treat Flag			ADSL
54	ADAE	1	STUDYID	text	15	Study Identifier	1		ADSL
55	ADAE	2	USUBJID	text	25	Unique Subject Identifier	2		ADSL
56	ADAE	3	SITEID	text	7	Study Site Identifier			ADSL
57	ADAE	4	COUNTRY	text	3	Country			ADSL
58	ADAE	5	AESEQ	integer	8	Sequence Number			AE
59	ADAE	6	AGE	integer	8	Unique Subject Identifier			ADSL
60	ADAE	7	AGEGR1N	integer	8	Pooled Age Group 1 (N)			ADSL
61	ADAE	8	AGEGR1	text	20	Pooled Age Group 1			ADSL
62	ADAE	9	SEX	text	1	Sex			ADSL
63	ADAE	10	TRTAN	integer	8	Actual Treatment (N)			ADSL.TRT01AN
64	ADAE	11	TRTA	text	40	Actual Treatment			ADSL.TRT01A

	COMMENTOID	DISPLAYFORMAT	COMPUTATIONMETHOD	CODELISTNAME	MANDATORY	ROLE	ROLECODELIST
33			AGEGR1	AGEGR1	Yes		
34				SEX	Yes		
35				RACE	Yes		
36				ARMN	Yes		
37			TRTP	ARM	No		
38	COM.SEECODELIST			ADEF.PARAMCD	No		
39	COM.SEECODELIST			ADEF.PARAM	No		
40				VISIT	No		
41			COMP.SEECODELIST	AVISITN	No		
42			ABLFL	FLAGS	No		
43					No		
44				VISITNUM	No		
45		date9.	ADT		No		
46			ADY		No		
47			COMP.SEEVALUELEVEL		No		
48			COMP.SEEVALUELEVEL		No		
49			BASE		No		
50			CHANGEFROMBASELINE		No		
51			COMP.SEEVALUELEVEL	YN	No		
52			COMP.SEEVALUELEVEL		No		
53				FLAGS	No		
54					Yes	Identifier	ROLECODE
55					Yes	Identifier	ROLECODE
56					Yes	Record Qualifier	ROLECODE
57				COUNTRY	Yes	Record Qualifier	ROLECODE
58					Yes		
59					Yes	Identifier	ROLECODE
60			COMP.SEECODELIST		Yes		
61			AGEGR1	AGEGR1	Yes		
62					Yes	Synonym Qualifier	ROLECODE
63					Yes	Timing	ROLECODE
64					No	Timing	ROLECODE

	DOMAIN	VARNUM	VARIABLE	TYPE	LENGTH	LABEL	KEYSEQUENCE	SIGNIFICANTDIGITS	ORIGIN
65	ADAE	12	AETERM	text	200	Reported Term for the Adverse Event			AE
66	ADAE	13	AEDECOD	text	200	Dictionary-Derived Term	4		AE
67	ADAE	14	AEBODSYS	text	200	Body System or Organ Class			AE
68	ADAE	15	ASTDT	integer	8	Start Date/Time of Adverse Events	5		Derived
69	ADAE	16	AENDT	integer	8	End Date/Time of Adverse Events			Derived
70	ADAE	17	ASTDY	integer	8	Study Day of Start of Adverse Event			Derived
71	ADAE	18	AENDY	integer	8	Study Day of End of Adverse Event			Derived
72	ADAE	19	AESEV	text	40	Severity/Intensity			AE
73	ADAE	20	AESEVN	integer	8	Severity/Intensity (N)			Derived
74	ADAE	21	AESER	text	40	Serious Event			AE
75	ADAE	22	AEACN	text	40	Action Taken with Study Treatment			AE
76	ADAE	23	AEREL	text	40	Causality			AE
77	ADAE	24	AERELN	integer	8	Causality (N)			Derived
78	ADAE	25	CQ01NAM	text	200	CQ 01 Name			Derived
79	ADAE	26	RELGR1	text	15	Pooled Causality Group 1			Derived
80	ADAE	27	RELGR1N	integer	8	Pooled Causality Group 1 (N)			Derived
81	ADAE	28	TRTEMFL	text	1	Treatment Emergent Flag			Derived
82	ADAE	29	SAFFL	text	1	Safety Population Flag			ADSL
83	ADTTE	1	STUDYID	text	15	Study Identifier	1		ADSL
84	ADTTE	2	USUBJID	text	25	Domain Abbreviation	2		ADSL
85	ADTTE	3	SITEID	text	7	Study Site Identifier			ADSL
86	ADTTE	4	COUNTRY	text	3	Country			ADSL
87	ADTTE	5	AGE	integer	8	Unique Subject Identifier			ADSL
88	ADTTE	6	AGEGR1N	integer	8	Pooled Age Group 1 (N)			ADSL
89	ADTTE	7	AGEGR1	text	20	Pooled Age Group 1			ADSL
90	ADTTE	8	SEX	text	1	Sex			ADSL
91	ADTTE	9	RACE	text	40	Race			ADSL
92	ADTTE	10	TRTPN	integer	8	Planned Treatment (N)			ADSL.TRTD1PN
93	ADTTE	11	TRTP	text	40	Planned Treatment			ADSL.TRTD1P
94	ADTTE	12	PARAMCD	text	8	Parameter Code	5		Assigned
95	ADTTE	13	PARAM	text	40	Parameter			Assigned
96	ADTTE	14	ADT	integer	8	Analysis Date			Derived
97	ADTTE	15	STARTDT	integer	8	Time to Event Origin Date for Subject			ADSL.RANDDT
98	ADTTE	16	AVAL	integer	8	Analysis Value			Derived
99	ADTTE	17	CNSR	integer	8	Censor			Derived
100	ADTTE	18	EVNTDESC	text	50	Event description			Derived
101	ADTTE	19	SRCDOM	text	8	Source Domain			Derived
102	ADTTE	20	SRCVAR	text	8	Source Variable			Derived
103	ADTTE	21	SRCSEQ	integer	8	Source Sequence Number			Derived
104	ADTTE	22	ITTFL	text	1	Intent-to-Treat Flag			ADSL

	J COMMENTOID	K DISPLAYFORMAT	L COMPUTATIONMETHOD	M CODELISTNAME	N MANDATORY	O ROLE	P ROLECODELIST
65					Yes	Topic	ROLECODE
66				MedDRA	Yes	Synonym Qualifier	ROLECODE
67				MedDRA	Yes	Record Qualifier	ROLECODE
68			ADAE.ASTDT		Yes	Timing	ROLECODE
69			ADAE.AENDT		Yes	Timing	ROLECODE
70			ADAE.ASTDY		No		
71			ADAE.AENDY		No		
72				AESEV	No	Record Qualifier	ROLECODE
73		1	COMP.SEECODELIST	AESEVN	No		
74				YN	No	Record Qualifier	ROLECODE
75				ACN	No	Record Qualifier	ROLECODE
76				AEREL	No	Record Qualifier	ROLECODE
77		1	COMP.SEECODELIST	AERELN	No		
78			COMP.CQ01NAME		No		
79			RELGR1	RELGR1	No		
80			COMP.SEECODELIST	RELGR1N	No		
81				FLAGS	No		
82				FLAGS	No	Record Qualifier	ROLECODE
83					Yes		
84					Yes		
85					Yes		
86				COUNTRY	Yes		
87					Yes		
88			COMP.SEECODELIST		Yes		
89			AGEGR1	AGEGR1	Yes		
90					Yes		
91				RACE	Yes		
92				ARMN	Yes		
93				ARM	No		
94				ADTTE.PARAMCD	No		
95				ADTTE.PARAM	No		
96					No		
97					No		
98			COMP.SEEVALUELEVEL		Yes		
99			COMP.SEEVALUELEVEL		Yes		
100			COMP.SEEVALUELEVEL		No		
101			ADTTE.SRCDOM		No		
102			ADTTE.SRCVAR		No		
103			ADTTE.SRCSEQ		No		
104				FLAGS	No		

Appendix C.4 - Parameter-Level Metadata

DOMAIN	VARIABLE	WHERECLAUSEOID	VALUEVAR	VARNUM	VALUENAME	TYPE	LENGTH	LABEL
ADEF	AVAL		PARAMCD	1	XPPAIN	integer	8	Pain Score
ADEF	CRIT1		PARAMCD	1	XPPAIN	text	51	Pain Score
ADEF	CRIT1FL		PARAMCD	1	XPPAIN	text	1	Pain Score
ADTTE	AVAL		PARAMCD	1	TTPNRELF	integer	8	Time to First Pain Relief
ADTTE	CNSR		PARAMCD	1	TTPNRELF	integer	8	Censor
ADTTE	EVNTDESC		PARAMCD	1	XPPAIN	text	50	Event or Censoring Description

SIGNIFICANTDIGITS	ORIGIN	COMMENTOID	DISPLAYFORMAT	COMPUTATIONMETHODOID	CODELISTNAME	MANDATORY	ROLE	ROLECODELIST
	XP.XPSTRESN		1.		PAINSCORE	No		
	Assigned	COM.CRIT1	$52.		ADEF.CRIT1	No		
	Derived		$1.	RESPONDER	YN	No		
	Derived		8.	TTPNRELF		No		
	Derived		8.	TTPNRELF.CNSR	CENSOR	No		
	Derived		$50	TTPNRELF.EVNTDESC	EVNTDESC	No		

Appendix C.5 - Computational Method Metadata

COMPUTATIONMETHODOID	LABEL	TYPE	COMPUTATIONMETHOD
BASE	Baseline value derivation	Computation	Last non-missing value of AVAL prior to first dose
CHANGEFROMBASELINE	Change from baseline calculation	Computation	AVAL-BASE
RESPONDER	Responder derivation	Computation	IF (.Z<CHG<=-2) THEN CRIT1FL='Y' ELSE CRIT1FL='N'
TTPNRELF	Time to first pain relief without worsening derivation	Computation	Derived from ADEF.ADY (where PARAMCD=XPPAIN and ADY>1) or ADAE.ASDY where CQ01NAM=PAIN EVENT. Value is the earliest day when ADEF.CHG<0 (for an event) or (ADEF.CHG>0 or ADAE.ASDY>0) (for subjects censored due to pain worsening) or the last record where ADEF.CHG=0 (for subjects censored due to no improvement or worsening by the time of the last assessment)
TTPNRELF.CNSR	Censoring algorithm for Time to First Pain Relief	Computation	if pain relief occurs before worsening, then CNSR=0; else if worsening from pain data, then CNSR=1; else if worsening from AE data, then CNSR=3; else if no relief and no worsening at time of last observation, then CNSR=4
TTPNRELF.EVNTDESC	Event descriptions for Time to First Pain Relief	Computation	if CNSR=0, then EVNTDESC=PAIN RELIEF; if CNSR=1, then EVNTDESC=PAIN WORSENING PRIOR TO RELIEF; if CNSR=2, then EVNTDESC=PAIN ADVERSE EVENT PRIOR TO RELIEF; if CNSR=3, then EVNTDESC=NO RELIEF AND NO WORSENING AT LAST ASSESSMENT
TRTSDT	Date of first exposure to treatment derivation	Computation	EX.EXSTDTC where EXSEQ=1, converted to a SAS date. See SAP for missing or partial value imputations
TRTEDT	Date of last exposure to treatment derivation	Computation	EX.EXENDTC converted to a SAS date for the last record in EX. See SAP for missing or partial value imputations
ITTFL	Intent-to-Treat flag derivation	Computation	If ADSL.RANDDT ^= missing then Y, else Null
SAFFL	Safety population flag derivation	Computation	If ADSL.TRTSDT^=missing then Y, else Null
RESPFL	Responder population flag derivation	Computation	If XP.VISIT='Month 6' and XPSTRESN represents at least a two-point improvement from the baseline record then Y, else N
RANDDT	Randomization date derivation	Computation	SUPPDM.QVAL where SUPPDM.QNAM=RANDDTC, converted to a SAS date

COMPUTATIONMETHODOID	LABEL	TYPE	COMPUTATIONMETHOD
ABLFL	Analysis baseline record flag	Computation	Last visit with a non-missing record prior to first dose
ADT	Analysis date calculation	Computation	XPDTC converted to a SAS date. See SAP for missing or partial value imputations
ADY	Analysis day calculation	Computation	ADT - ADSL.RANDDT + 1
ADAE.ASTDT	Analysis start date calculation	Computation	AESTDTC converted to a SAS date. See SAP for missing or partial value imputations
ADAE.AENDT	Analysis end date calculation	Computation	AEENDTC converted to a SAS date. See SAP for missing or partial value imputations
ADAE.ASTDY	Analysis start day calculation	Computation	ASTDT - ADSL.TRTSDT + 1
ADAE.AENDY	Analysis end day calculation	Computation	AENDT - ADSL.TRTEDT + 1
ADTTE.SRCVAR	SRCVAR algorithm	Computation	SRCVAR = ADEF if first event is a negative pain score or if worsening occurs before an improvement. SRCVAR = ADAE if ADAE.CQ01NAM is non-missing on a date prior to a pain improvement or worsening.
ADTTE.SRCSEQ	SRCSEQ algorithm	Computation	if SRCVAR=ADEF then SRCSEQ=ADEF.XPSEQ from the corresponding record, else if SRCVAR=ADAE then SRCSEQ=ADAE.AESEQ from the corresponding record.
COMP.SEECODELIST	See codelist	Computation	See codelist
COMP.SEEVALUELEVEL	See parameter value-level metadata	Computation	See parameter value-level metadata
RELGR1	AE relatedness derivation	Computation	If AEREL=NOT RELATED then NOT RELATED, Otherwise RELATED
AGEGR1	Age group derivation	Computation	If age<55 then "<55 years"; else if age>=55 then ">=55 YEARS"

Appendix C.6 – Where Clause Metadata

	A	B	C	D	E	F	G
1	WHERECLAUSEOID	SEQ	SOFTHARD	ITEMOID	COMPARATOR	VALUES	COMMENTOID
2	WC.LB.LBTESTCD.GLUC.LBCAT.CHEMISTRY	1	Soft	LB.LBCAT	EQ	CHEMISTRY	
3	WC.LB.LBTESTCD.GLUC.LBCAT.CHEMISTRY	2	Soft	LB.LBTESTCD	EQ	GLUC	
4	WC.LB.LBTESTCD.GLUC.LBCAT.URINALYSIS	1	Soft	LB.LBCAT	EQ	URINALYSIS	
5	WC.LB.LBTESTCD.GLUC.LBCAT.URINALYSIS	2	Soft	LB.LBTESTCD	EQ	GLUC	
6	WC.ITTFL	1	Soft	ADSL.ITTFL	EQ	Y	
7	WC.Table_14.2.1.1	1	Soft	ADEF.ITTFL	EQ	Y	
8	WC.Table_14.2.1.1	2	Soft	ADEF.ABLFL	NE	Y	
9	WC.Table_14.2.1.1	3	Soft	ADEF.PARAMCD	EQ	XPPAIN	
10	WC.Figure_14.2.1.1	1	Soft	ADTTE.ITTFL	EQ	Y	
11	WC.Figure_14.2.1.1	2	Soft	ADTTE.PARAMCD	EQ	TTPNRELF	
12	WC.Table_14.3.2.1	1	Soft	ADAE.TRTEMFL	EQ	Y	

Appendix C.7 – Comments Metadata

	A	B
1	COMMENTOID	COMMENT
2	COM.CRIT1	Assignment per the codelist
3	COM.CQ01NAME	if non-null then PAIN EVENT
4	COM.SEECODELIST	See codelist
5	COM.SEEVALUELEVEL	See parameter value-level metadata
6	COM.ADAE	Primarily derived from SDTM.AE
7	COM.ADSL	Derived from multiple SDTM domains
8	COM.ADEF	Primarily derived from SDTM.XP
9	COM.ADTTE	Primarily derived from ADaM.ADAE

Appendix C.8 - Codelist Metadata

CODELISTNAME	RANK	CODEDVALUE	TRANSLATED	TYPE	CODELISTDICTIONARY	CODELISTVERSION	ORDERNUMBER
ACN	1	DOSE INCREASED	DOSE INCREASED	text			
ACN	2	DOSE NOT CHANGED	DOSE NOT CHANGED	text			
ACN	3	DOSE REDUCED	DOSE REDUCED	text			
ACN	4	DRUG INTERRUPTED	DRUG INTERRUPTED	text			
ACN	5	DRUG WITHDRAWN	DRUG WITHDRAWN	text			
ACN	6	NOT APPLICABLE	NA	text			
ACN	7	UNKNOWN	U; Unknown	text			
AEREL	1	NOT RELATED	NOT RELATED	text			
AEREL	2	POSSIBLY RELATED	POSSIBLY RELATED	text			
AEREL	3	PROBABLY RELATED	PROBABLY RELATED	text			
AESEV	1	MILD	Grade 1, 1	text			
AESEV	2	MODERATE	Grade 2, 2	text			
AESEV	3	SEVERE	Grade 3, 3	text			
AGEU	1	YEARS	YEARS	text			
ARM	1	Analgezia HCL 30 mg	Analgezia HCL 30 mg	text			
ARM	2	Placebo	Placebo	text			
ARMN	1	1	Analgezia HCL 30 mg	integer			
ARMN	2	0	Placebo	integer			
ARMCD	1	ALG123	Analgezia HCL 30 mg	text			
ARMCD	2	PLACEBO	Placebo	text			
COUNTRY	1	USA	UNITED STATES	text			
SEX	1	F	FEMALE	text			
SEX	2	M	MALE	text			
SEX	3	U	UNKNOWN	text			
VISIT	1	Baseline	Baseline	text			
VISIT	2	Month 3	Month 3	text			
VISIT	3	Month 6	Month 6	text			
VISITNUM	1		0 Baseline	text			
VISITNUM	2		1 Month 3	text			
VISITNUM	3		2 Month 6	text			
YN	2	Y	YES	text			
YN	1	N	NO	text			
TRTG1PN	2	1	Analgezia HCL 30 mg	integer			

CODELISTNAME	RANK	CODEDVALUE	TRANSLATED	TYPE	CODELISTDICTIONARY	CODELISTVERSION	ORDERNUMBER
TRTG1PN	1	0	Placebo	integer			
PAINSCORE	0	0	None	integer			
PAINSCORE	1	1	Mild	integer			
PAINSCORE	2	2	Moderate	integer			
PAINSCORE	3	3	Severe	integer			
AVISITN	1	0	Baseline	integer			
AVISITN	2	3	Month 3	integer			
AVISITN	3	6	Month 6	integer			
AERELN	1	0	NOT RELATED	integer			
AERELN	2	1	POSSIBLY RELATED	integer			
AERELN	3	2	PROBABLY RELATED	integer			
AESEVN	1	1	Mid	integer			
AESEVN	2	2	Moderate	integer			
AESEVN	3	3	Severe	integer			
RELGR1N	1	0	Not related	integer			
RELGR1N	1	1	Related	integer			
EVNTDESC	1	PAIN RELIEF	Pain relief	text			
EVNTDESC	2	PAIN WORSENING PRIOR TO RELIEF	Pain worsening prior to relief	text			
EVNTDESC	3	PAIN ADVERSE EVENT PRIOR TO RELIEF	Pain adverse event prior to relief	text			
EVNTDESC	4	NO RELIEF AND NO WORSENING AT LAST ASSESSM	No relief and no worsening at last assessment	text			
RELGR1	1	Not related	Not related	text			
RELGR1	2	Related	Related	text			
CRITVALUES	1	Pain improvement from baseline of at least 2 points	Pain improvement from baseline of at least 2 points	text			
CENSOR	1	0	Pain relief	integer			
CENSOR	2	1	Pain worsening prior to relief	integer			
CENSOR	3	2	Pain adverse event prior to relief	integer			
CENSOR	4	3	No relief and no worsening at last assessment	integer			
ADEF PARAM	1	Pain Score	Pain Score	text			
ADEF PARAMCD	1	XPPAIN	Pain Score	text			

CODELISTNAME	RANK	CODEDVALUE	TRANSLATED	TYPE	CODELISTDICTIONARY	CODELISTVERSION	ORDERNUMBER
ADTTE PARAM	1	Time to first pain relief	Time to first pain relief	text			
ADTTE PARAMCD	1	TTPNRELF	Time to first pain relief	text			
ADEF CRIT1	1	Pain improvement from baseline of at least 2 points	Pain improvement from baseline of at least 2 points	text			
FLAGS	1	Y	Yes	text			
AGEGR1N	1	1	<55 Years	integer			
AGEGR1N	2	2	>=55 Years	integer			
AGEGR1	1	<55 YEARS	<55 Years	text			
AGEGR1	2	>=55 YEARS	>=55 Years	text			
RACE	1	BLACK OR AFRICAN AMERICAN	BLACK OR AFRICAN AMERICAN	text			
RACE	2	ASIAN	ASIAN	text			
RACE	3	WHITE	WHITE	text			

Appendix C.9 - Analysis Results Metadata

	A	B	C	D	E	F	G	H	I
1	DISPLAYID	DISPLAYNAME	RESULTNAME	REASON	PURPOSE	PARAMCD	ANALYSISVARIABLES	ANALYSISDATASET	WHERECLAUSEOID
2	Table_14.1.1	Summary of Demographics (ITT Population)	Summary of Demographics	SPECIFIED IN SAP	Comparisons of baseline characteristics by treatment group		AGE, AGEGR1, SEX, RACE	ADSL	WC.ITTFL
3	Table_14.2.1	Responder Analysis by Visit	Responder Analysis by Visit	SPECIFIED IN SAP	Primary efficacy endpoint	XPPAIN	CRIT1FL	ADEF	WC.Table_14.2.1.1
4	Table_14.2.2	Mean Pain Assessments and Changes from Baseline	Summary of Pain Relief	SPECIFIED IN SAP	Secondary efficacy endpoint	XPPAIN	AVAL, CHG	ADEF	WC.Table_14.2.1.1
5	Figure_14.2.1	Time to Pain Relief Analysis by Treatment Group	Time to Pain Relief Kaplan Meier Analysis	SPECIFIED IN SAP	Secondary efficacy endpoint	TTPNRELF	AVAL	ADTTE	WC.Figure_14.2.1.1
6	Figure_14.2.1	Time to Pain Relief Analysis by Treatment Group	Time to Pain Relief Hazard Ratio Derivation	SPECIFIED IN SAP	Secondary efficacy endpoint	TTPNRELF	AVAL	ADTTE	WC.Figure_14.2.1.1
7	Table_14.3.2.1	Incidence Rates of TEAEs by System Organ Class and Preferred Term	Incidence Rates of TEAEs by System Organ Class and Preferred Term	SPECIFIED IN SAP	Secondary safety endpoint	N/A	AEBODSYS, AEDECOD	ADAE	WC.Table_14.3.2.1

	J	K	L	M	N
1	DOCUMENTATION	REFLEAFID	CONTEXT	PROGRAMMINGCODE	PROGRAMLEAFID
2	Rates and chi-squared tests of categorical demographic variables	SAP_Section_9.1.1	SAS Version 9.4	PROC FREQ DATA=ADSL; where ittfl='Y'; tables trt01pn * (agegr1 sex race) / cmh; run;	
3	Chi-squared test of responder rates by treatment group	SAP_Section_9.2.1	SAS Version 9.4	proc freq data = adef; by crit1 avisitn; where ittfl='Y'; tables trtpn * crit1fl / chisq; run;	
4	ANCOVA analysis of the treatment group effect with a baseline covariate, by visit	SAP_Section_9.2.1	SAS Version 9.4	proc glm data = adef; by avisitn avisit; class trtpn; model aval = trtpn base; run;	
5	Kaplan-Meier estimates and log-rank test for the treatment effect	SAP_Section_9.2.2	SAS Version 9.4	proc lifetest data = adtte plots=s; where ittfl = 'Y' and paramcd = 'TTPNRELF'; id usubjid; strata trtpn; time aval*cnsr (1 2 3); test trtpn; run;	
6	Cox proportional hazards model used to derive the hazard ratio for the treatment group effect	SAP_Section_9.2.2	SAS Version 9.4	proc phreg data = adtte; model aval*cnsr(1, 2, 3) = trtpn; run;	
7	Incidence rates of TEAEs by System Organ Class and Preferred Term	SAP_Section_9.3.1	SAS Version 9.4		ae-t14-3-02

Appendix C.10 - External Links Metadata

	LeafID	LeafRelPath	LeafPageRef	LeafPageRefType	Title	SupplementalDoc	AnnotatedCRF
1	LeafID	LeafRelPath	LeafPageRef	LeafPageRefType	Title	SupplementalDoc	AnnotatedCRF
2	ADRG	reviewersguide.pdf			Analysis Data Reviewer's Guide	Y	
3	SAP_Section_9.1.1	.\.\.\.\53-clin-stud-rep\535-rep-effic-safety-stud\5351-stud-rep-contr\studyxyz123\studyxyz123_csr.pdf	SAP_SEC_9.1.1	NamedDestination	SAP Section 9.1.1		
4	SAP_Section_9.2.1	.\.\.\.\53-clin-stud-rep\535-rep-effic-safety-stud\5351-stud-rep-contr\studyxyz123\studyxyz123_csr.pdf	SAP_SEC_9.2.1	NamedDestination	SAP Section 9.2.1		
5	SAP_Section_9.2.2	.\.\.\.\53-clin-stud-rep\535-rep-effic-safety-stud\5351-stud-rep-contr\studyxyz123\studyxyz123_csr.pdf	SAP_SEC_9.2.2	NamedDestination	SAP Section 9.2.2		
6	SAP_Section_9.3.1	.\.\.\.\53-clin-stud-rep\535-rep-effic-safety-stud\5351-stud-rep-contr\studyxyz123\studyxyz123_csr.pdf	SAP_SEC_9.3.1	NamedDestination	SAP Section 9.3.1		
7	Table_14.1.1	.\.\.\.\53-clin-stud-rep\535-rep-effic-safety-stud\5351-stud-rep-contr\studyxyz123\studyxyz123_csr.pdf	302	PhysicalRef	Table 14.1.1		
8	Table_14.2.1	.\.\.\.\53-clin-stud-rep\535-rep-effic-safety-stud\5351-stud-rep-contr\studyxyz123\studyxyz123_csr.pdf	352	PhysicalRef	Table 14.2.1		
9	Table_14.2.2	.\.\.\.\53-clin-stud-rep\535-rep-effic-safety-stud\5351-stud-rep-contr\studyxyz123\studyxyz123_csr.pdf	353	PhysicalRef	Table 14.2.2		
10	Figure_14.2.1	.\.\.\.\53-clin-stud-rep\535-rep-effic-safety-stud\5351-stud-rep-contr\studyxyz123\studyxyz123_csr.pdf	354	PhysicalRef	Figure 14.2.1		
11	ae-t14-3-02	ae-t14-3-02.sas			AE Analysis Program		

Appendix D: %run_p21v SAS macro

In this appendix, we present the %run_p21v SAS macro that is used in Chapter 9 to validate our CDISC implementations using Pinnacle 21 software in batch mode. The program is annotated with numbered bullets for additional commentary that appears at the end of this appendix.

%run_p21v SAS Macro

```
*-------------------------------------------------------------------------;
* Run the Pinnacle 21 Community Validator in batch mode                   ;
*  TYPE= The standard being validated (SDTM, ADaM, Define, SEND, or
*    Custom)
*  SOURCES= Path of the directory that contains the SAS datasets or
*    define file
*  P21Path= Path of the local OpenCDISC Validator installation
*  ValidatorJAR= Subdirectory and name of the JAR file
*
*  FILES=       Defaults to all .xpt files in the directory. Alternatively,
*               one specific file could be specified or a wildcard could
*               be used on a partial name (e.g. LB*.xpt)
*
*  REPORTFNAME= Name of the output Excel file that contains the validation
*               report.  Can be left blank to have automated naming occur.
*
*  CONFIG=      Subdirectory and name of the configuration file to be
*               used for validation.  Should correspond to one of the
*               files in the CONFIG sub-directory.
*
*  DEFINE=Y     Choose Y, y, or 1 if a DEFINE.XML file exists (in the
*               same directory as the data sets) for cross-validation.
*               Do NOT choose DEFINE=Y/y/1 if you wish to validate a
*               define file.
*
*  CTDATADATE= Specify the version of the controlled terminology XML file
*               used to check against.  Since the file locations and names
*               are standardized, all that is needed is the date in
*               yyyy-mm-dd format. There are many different versions from
*               which to choose and these are updated frequently.  Refer
*               to your P21 software and/or the COMPONENTS\CONFIG\DATA\CDISC
*               directory for CDISC controlled terminology.
*
*  MAKE_BAT=N  Specify Y if a .BAT file is desired.  The .BAT file can
*               be submitted/typed at a command prompt (without the .BAT
*               extension) to initiate the process of having a validation
*               report generated (without having to resubmit this macro)
*-------------------------------------------------------------------------;
%macro run_p21v(type= ,
                sources= ,
                ctdatadate= ,
                p21path=C:\Software\pinnacle21-community,        ❶
                validatorjar=\components\lib\validator-cli-2.1.0.jar,  ❷
                files=*.xpt,
                reportfname= ,
                config= ,
```

```
                         define=Y,
                         make_bat=N
                         );

      ** set the appropriate system options;
      options symbolgen xsync noxwait ;

      ** specify the output report path;
      %let reportpath=&sources;

      ** ensure proper capitalization in case of case sensitivity;
      %if %upcase(&type)=ADAM %then
        %let type=ADaM;

      ** specify the name(s) of the validation reports;           ❸
      %if &reportfname^= %then
        %let reportfile=&reportfname;
      %else %if %upcase(&type)=SDTM %then
        %let reportfile=sdtm-validation-report-
      &sysdate.T%sysfunc(tranwrd(&systime,:,-)).xlsx;
        %else %if %upcase(&type)=ADAM %then
        %let reportfile=adam-validation-report-
      &sysdate.T%sysfunc(tranwrd(&systime,:,-)).xlsx;
        %else
        %let reportfile=p21-community-validation-report-
      &sysdate.T%sysfunc(tranwrd(&systime,:,-)).xlsx;
        ;

      ** specify the name(s) of the configuration file (if missing);
      ** (note that the names of the configuration files can change with new
      CDISC
      ** IGs and software releases so maintenance of this is required!!);  ❹
      %if &config= %then
        %do;
          %let configpath=&p21path\components\config;
          %if %upcase(&type)=SDTM %then
            %let config=&configpath\SDTM 3.2.xml;
          %else %if %upcase(&type)=ADAM %then
            %let config=&configpath\ADaM 1.0.xml;
          %else %if %upcase(&type)=DEFINE %then
            %do;
              %let config=&configpath\Define.xml.xml;            ❺
              %** ensure that [&]DEFINE=N in this case;
              %let define=N;
            %end;
        %end;

      %** cross-check against a define file?  if so, assume it exists in the
      same directory and is
      %** named simply define.xml ;
      %** if validating a define, then use [&]files as the define file name;
      %if %upcase(&type)=DEFINE %then
        %let config_define=%str(-config:define="&sources\&files");
      %else %if %upcase(&define)=Y or &define=1 %then
        %let config_define=%str(-config:define="&sources\define.xml");
      %else
        %let config_define= ;
        ;

      %** if ctdatadate is non-missing, then check controlled terminology
      against the corresponding file;                     ❻
      %if "&ctdatadate"^= "" %then
        %do;
          %let tdata=&p21path\components\config\data\CDISC\&type\&ctdatadate\
```

```
&type Terminology.odm.xml;
        %let config_codelists=%str(-config:codelists="&ctdata");
      %end;
    %else
      %let config_codelists= ;
    ;

    %put submitting this command: ;
    %put java -jar "&p21path\&validatorjar" -type=&type -
source="&sources\&files" -config="&config"
        &config_codelists &config_define -report="&reportpath\&reportfile"
-report:overwrite="yes" ;

    * run the report;
    x java -jar "&p21path\&validatorjar" -type=&type -
source="&sources\&files" -config="&config"
      &config_codelists &config_define -report="&reportpath\&reportfile" -
report:overwrite="yes" ; *> &sources\run_p21v_log.txt;

    %if &make_bat=Y %then
      %do;
        * send the command to a bat file;
        data _null_;

            file "&sources\submit_p21_&type._job.bat" ; ❼
            put 'java -jar "' "&p21path\&validatorjar" '"' " -type=&type -
source=" '"' "&sources\&files" '"' -config="' "&config" '"'
              " %bquote(&config_codelists) %bquote(&config_define) -
report=" '"' "&reportpath\&reportfile" '"' ' -report:overwrite="yes" ';
        run;
      %end;

%mend run_p21v;
```

Additional SAS Code Comments

❶ The root path of the Pinnacle 21 Community software installation will need to be customized for your environment and should be done here.

❷ New versions of the software will have new names for the JAR files that correspond to the software version number. So these will need to be maintained when the software is updated.

❸ A static file name can be provided via the REPORTFNAME macro parameter. Otherwise, the macro will create a dynamic name that uses the system date/time in the file name. This is similar to how the software operates in the GUI mode. For example, if validating an SDTM file on August 27, 2016 at 8:27 in the morning, the name of the validation report will be sdtm-validation-report-27AUG16T08-27.xlsx. If a static name is not provided and the validation being done is not for SDTM or ADaM, then the validation report will be named p21-community-validation-report-27AUG16T08-27.xlsx

❹ Similar to the JAR files, the names for configuration files can change with new releases of, for example, implementation guides. Note also that most standards now have two versions of the configuration file—one for FDA rules and one for PMDA rules. For example, the SDTM IG version 3.1.3 the file names are "SDTM 3.1.3 (FDA).xml" and "SDTM 3.1.3 (PMDA).xml".

❺ Note that if you are validating a define file, there should not be any cross-checking with the define file because you can't cross check a file with itself. So whatever value is provided for the DEFINE macro parameter is ignored and this is set to N if the TYPE parameter is set to DEFINE.

❻ The controlled terminology files have a standard folder structure and naming convention. If we already know the standard being implemented (e.g. SDTM, ADaM), then the date corresponding to the release of the controlled terminology used for your implementation can be provided in the CTDATADATE parameter in YYYY-MM-DD format. This should correspond to the folder names in the ..\components\config\data\CDISC\[standard type]\ subdirectories.

❼ Rather than repeatedly running this macro for the iterative validation process, you can instead have the macro create a .BAT file that you can either manually run at the command line or have SAS run via the X command. If doing this, however, you may want to supply a static filename for the REPORTFNAME macro parameter rather than having a dynamic date/time-based name specified at the time of the macro execution that may become quickly outdated.

Index

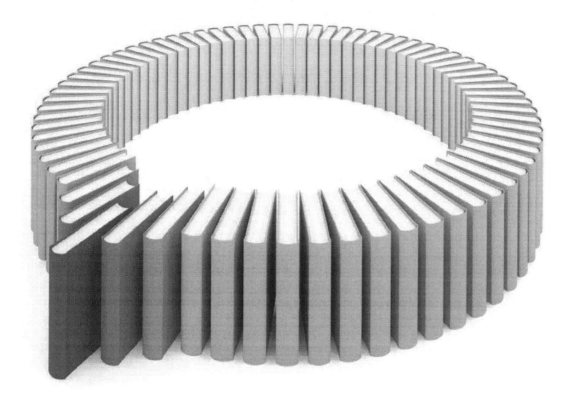

Gain Greater Insight into Your SAS® Software with SAS Books.

Discover all that you need on your journey to knowledge and empowerment.

support.sas.com/bookstore
for additional books and resources.

Made in the USA
Middletown, DE
15 December 2016